RARITY

Population and Community Biology Series

Principal Editor

M. B. Usher
Chief Scientific Adviser, Scottish Natural Heritage, UK

Editors

D. L. DeAngelis
Senior Scientist, Environmental Sciences Division
Oak Ridge National Laboratory, USA

R. L. Kitching
Professor, Department of Ecosystem Management,
University of New England, Australia

The study of both populations and communities is central to the science of ecology. This series of books explore many facets of population biology and the processes that determine the structure and dynamics of communities. Although individual authors are given freedom to develop their subjects in their own way these books are scientifically rigorous and a quantitative approach to analysing population and community phenomena is often used.

Already published

RARITY

Kevin J. Gaston

Department of Entomology, The Natural History Museum, London, UK

Learning Resources
Centre

CHAPMAN & HALL.

London · Glasgow · Weinheim · New York · Tokyo · Melbourne · Madras

Published by Chapman & Hall, 2-6 Boundary Row, London SE1 8HN, UK

Chapman & Hall, 2-6 Boundary Row, London SE1 8HN, UK

Blackie Academic & Professional, Wester Cleddens Road, Bishopbriggs, Glasgow G64 2NZ, UK

Chapman & Hall GmbH, Pappelallee 3, 69469 Weinheim, Germany
Chapman & Hall Inc., One Penn Plaza, 41st Floor, NY 10119, USA
Chapman & Hall Japan, Thomson Publishing Japan, Hirakawacho Nemoto Building, 6F, 1-7-11 Hirakawa-cho, Chiyoda-ku, Tokyo 102, Japan

Chapman & Hall Australia, Thomas Nelson Australia, 102 Dodds Street, South Melbourne, Victoria 3205, Australia

Chapman & Hall India, R. Seshadri, 32 Second Main Road, CIT East, Madras 600 035, India

First edition 1994

© 1994 Kevin J. Gaston

Typeset in 10/12pt Times by ROM-Data, Falmouth, Cornwall, England
Printed in England by Clays Ltd, St Ives plc

ISBN 0 412 47500 6 (HB)   0 412 47150 3 (PB)

∞ Printed on permanent acid-free text paper, manufactured in accordance with ANSI/NISO Z39.48-1992 and ANSI/NISO Z39.48-1984 (Permanence of Paper).

Contents

vi   Contents

Preface

To say you are writing about rarity is to invite two kinds of response. Either one provokes a discussion of what rarity is, or some comment on the complexity of the subject. The objective of this book is to explore the nature of rarity, its complexity if you like, from one particular perspective on what rarity is. Primarily, it is an opportunity to review, to synthesize, and to question. The book is an attempt to draw together a vast body of literature, to extract from it some general principles, and to raise question marks over areas the foundations of which appear to be either absent or crumbling.

A perusal of prefaces suggests that they often dwell as long upon what a book is not about, as upon what it does concern. True to such a tradition, I should state that this is specifically not a book about conservation, although in some quarters anything about rarity is viewed as something about conservation. Nor does it contain more than a passing reference to the undoubtedly important issues of the role of genetics in rarity.

Examples have been drawn from a wide variety of taxa. They are, nonetheless, somewhat depauperate in cases from marine systems. In part this bias results from the unevenness of my familiarity with the literature, in part it perhaps also reflects differences in the questions asked and approaches to the study of communities and assemblages in terrestrial and marine systems. Undoubtedly there will be instances in which I have missed many of the best examples. One of the problems of writing a volume on rarity is that these examples often come from studies which were not explicitly of this subject.

Some of the viewpoints expressed herein will be found to be at odds with statements I have made in the past. This is, by and large, a product of exposure to a wider range of literature brought about in writing this volume, and by taking a longer and harder look at some topics. I trust the reader will not begrudge such inconsistencies.

While the book is hopefully open to being 'picked at' rather than read from cover to cover, the reader is advised to become acquainted with the opening chapter first. Without an understanding of the way in which several terms are used, most especially 'rarity' itself, much of the rest of the book may be less intelligible. Certainly this will be true of the annotations to some of the figures.

A great many people have facilitated the development of my thoughts on the subject of rarity, and have on occasion kindly sought out or drawn my attention to pieces of information that might otherwise have eluded me. Most especially I would thank Tim Blackburn, Nigel Collar, Phil DeVries, Ian Gauld, Peter Hammond, Bill Kunin, John Lawton, Brian McArdle, Laurence

Mound, Nigel Stork, Michael Usher, Phil Warren, Paul Williams and Mark Williamson. The staff of the libraries of The Natural History Museum endured and answered numerous questions, while, with equal good humour, Natasha Loder shouldered some of the burden of extracting data and drawing figures. Sian Gaston, John Lawton, Michael Usher, Phil Warren, Paul Williams and Mark Williamson kindly read and commented on drafts of some or all of the chapters, undertakings for which I am deeply grateful. Bob Carling and Clem Earle at Chapman & Hall have both been a pleasure to work with. Finally, I am indebted to Sian for maintaining her enthusiasm when mine ebbed.

Introduction

The rare holds a curious fascination for humankind. From first editions to matchboxes, and from garden gnomes to cars, they are avidly collected and listed, categorized and counted, bought and sold. The stories of the lengths to which ornithologists will go to observe a species new to 'their patch' are apocryphal. Pitifully sad are the lengths to which collectors will go to obtain representatives of a plant the population of which is on the brink of extinction.

This enthusiasm for the rare has led to a widespread perception that they are special in some way which transcends their rarity. Science cannot escape the subjective, and similar feelings commonly surround those plant and animal species fortunate or unfortunate enough to be labelled as rare. But, what is a rare species? Does it really differ from those species which are not rare? If so, in what way? These are some of the questions which in eight chapters this book sets out to address.

In general, the material in the following pages becomes progressively complex through the book. Unsurprisingly, the starting point (Chapter 1) is a consideration of the definition of rarity, and some of the factors which serve to complicate or thwart unambiguous terminology. For reasons discussed in Chapter 1, I have chosen to regard rarity simply as the state of having a low abundance and/or a small range size. Chapter 2 moves on to explore the measurement of rarity on the basis of these two variables. Emphasis is laid on the variety of ways in which abundance and range size can be expressed, and the need to understand the limitations of the different approaches. Our understanding of inter-specific abundance and range size distributions is explored.

Although it is often convenient to treat them separately, abundance and range size are not independent. The objective of Chapter 3 is to explore their relationship. It divides broadly into two parts. The first is concerned with intra-specific interactions, which although not explicitly concerned with rarity provides some background for the second section. This addresses both the patterns of and the explanations for inter-specific interactions.

Chapters 4 and 5 are in some sense a pair, dwelling respectively on the spatial and the temporal dynamics of rarity. There are strong parallels in some of the concerns of both, although the two topics are treated somewhat differently. Chapter 4 asks whether there is any tendency for species which are rare in one area also to be rare in another. Answers depend in part on spatial scale and ultimately necessitate an understanding of the abundance structure of geographic ranges. Chapter 5 asks whether species which are rare at present are likely to have been rare in the past and will also be rare in the future. As well as leading to issues of scale, this raises matters such as the relationship between rarity and persistence, the population dynamics of rarity, and the trajectories by which species become extinct.

A full understanding of the dynamics of rarity, in time and space, necessitates tackling the rather tortuous subject of what causes species to be rare. This provides the topic of Chapter 6. The central thesis explored is that there is no general theory of the causes of rarity. Rather, rarity is generated by the same processes which act, albeit less severely, upon the abundances and range sizes of common and widespread species. Two broad groups of factors are recognized as generating low abundances and/or small range sizes, environmental variables (abiotic and biotic) and colonization abilities (dispersal and establishment).

The logical conclusion from arguing that there is no general theory of rarity is that there is no panacea for dealing with those species whose rarity is regarded as a problem. Chapter 7 is given over to a consideration of the most important applied aspect of the study of rarity, that of conservation. The conservation of rare species is tied foremost to the view that a central objective of conservation is the prevention or limitation of the extinction of species. Thus, the factors affecting the likelihood of extinction are examined. Accepting its artificiality, a distinction is made between site-based and species-based approaches to conservation, and the role of rare species in each is considered.

Chapter 8 serves as a conclusion in that it draws together some of the common themes of the preceding chapters, and asks the question 'Where next?'. This divides broadly into two sets of issues, how the study of rarity can be improved and the primary issues which remain to be resolved.

1 What is rarity?

Discussions will soon be at cross-purposes if it is not clearly recognised that our concepts of what is rare will depend on the scale of our individual experience and on the range or narrowness of our special interests.

J.L. Harper (1981)

Rarity is one of those concepts that suffuses our culture: it defies precise definition and when used by the scientist it is often given a spurious accuracy to satisfy our need for precision.

V.H. Heywood (1988)

Population and community biology have repeatedly been plagued with problems of definition. The meaning of such fundamental terms and concepts as competition, density dependence, carrying capacity, niche, stability and community have all, and often for prolonged periods, been debated (e.g. Pimm, 1984; Giller and Gee, 1987; Murray, 1987; Dhondt, 1988; Keddy, 1989; Schoener, 1989). Such discussion of the use and interpretation both of words and concepts may, on occasion, degenerate into arguments over semantics. However, for any area of study, clarity of terminology is essential to the establishment of a rigorous framework in which both theoretical and empirical work can be placed. With regard to the study of rarity no general theory and no such framework presently exist. At the outset, therefore, we would do well to be clear about what we mean by rarity and rare species.

In this chapter the view of rarity upon which this book is founded is established, and its relation to a number of issues is explored. The first section identifies the variables by which rarity is defined, and the second how, on the basis of those variables, species which are rare can be recognized. The third and fourth sections address the relationships of spatial scale and endemism to concepts of rarity, and the following section considers how to delineate the assemblage with respect to which species are viewed as rare. The penultimate section is concerned with the interaction between rarity and vulnerability, and the concluding section outlines some further features of how rarity will be used in this book.

1.1 THE VARIABLES USED TO DEFINE RARITY

The term 'rare' has a variety of meanings in common usage (Table 1.1; Harper, 1981). However, in the context of population and community biology it is, by and large, used in a somewhat more constrained sense. Rare species are regarded as those having low abundance and/or small ranges. What is meant

2   What is rarity?

Table 1.1 Extracts from The Shorter Oxford English Dictionary (1983) defining the term 'rare'

1 Having the constituent particles not closely packed together.
2 (a) Having the component parts widely set; of open construction; in open order.
(b) Thinly attended or populated.
3 Placed or stationed at wide intervals; standing or keeping far apart.
4 Few in number and widely separated from each other (in space or time); forming a small and scattered class.
5 Of a kind, class, or description seldom found, met with, occurring; unusual, uncommon, exceptional.
6 Unusual in respect of some good quality; remarkably good or fine.

by this is unfortunately rather infrequently expressed. Usher (1986a) neatly captures the situation in describing rarity as 'an intuitive concept'. Indeed, while the magnitude of abundance and range size may serve as some lowest common denominator, there remains a bewildering diversity of viewpoints on the limits to what constitutes rarity in population and community biology (Mayr, 1963; Drury, 1974, 1980; Harper, 1981; Margules and Usher, 1981; Rabinowitz, 1981a; Reveal, 1981; Main, 1982; Cody, 1986; Fiedler, 1986; Rabinowitz *et al.*, 1986; Soulé, 1986; Usher, 1986a, b; Heywood, 1988; Ferrar, 1989; Hanski, 1991a; Fiedler and Ahouse, 1992; McCoy and Mushinsky, 1992; Reed, 1992). Habitat specificity, taxonomic distinctness, and persistence through ecological or evolutionary time, have variously been regarded either as additional variables by which rare species can be identified, or as restrictions upon which species with low abundances or small range sizes are regarded as rare. In many instances the resultant definitions serve to prejudge the causes of the phenomenon. Some of these definitions may be reconciled but many may not.

There are various possible ways to move forward from the present confused position. One approach would be to seek further variables that may be used to define rare species. However, this seems likely to generate more, rather than less, confusion. Here I have chosen to regard rarity as simply being the state of having a low abundance and/or a small range size. I seek to limit it on the basis of no additional variables. This follows Reveal's (1981) statement that '. . . rarity is merely the current status of an extant organism which, by any combination of biological or physical factors, is restricted either in numbers or area to a level that is demonstrably less than the majority of other organisms of comparable taxonomic entities.' In the rest of this chapter I consider how one might best go about deciding, on the basis of abundance and range size, precisely which species are rare.

1.2 CONTINUOUS AND DISCONTINUOUS MEASURES

Both abundance and range size are essentially continuous variables (although an abundance expressed as numbers of individuals can, of course, only have integer values). Moreover, for any natural assemblage it is unusual to find that species separate into discrete groups with abundances or range sizes of distinctly different magnitudes. Typically, an assemblage will comprise species with a range of different abundances and levels of spatial occurrence (for some exceptions see Chapter 2). Against this background it makes some sense to regard rarity as a continuous variable. That is, all species in essence are treated as rare, but some (those with low abundances or small ranges) are rarer than others (those with high abundances or large ranges). Thus, rarity would effectively be the inverse of the magnitude of abundance, of range size, or of some combination of the two.

It is difficult to find any compelling objections to regarding rarity as a continuous variable. Indeed, the term 'rare' is frequently used as a synonym for low abundance or small range, and as the converse of common or widespread, without any attempt to place hard limits on where rarity ends and commonness begins. Thus, Brown (1984) uses 'rare' to mean an extremely low density ('restricted' or 'local' was used to describe an extremely small spatial distribution), and Schoener (1987) uses it to mean occurrence in relatively few censuses and/or at relatively low abundances.

The chief arguments against regarding rarity as a continuous variable are twofold. First, that it essentially equates the study of rarity with the study of the whole of population biology, and de-emphasizes any claim rare species may have to be of special interest. Second, because for legal and conservation purposes species often need to be categorized as rare or otherwise, a more pragmatic approach is often desirable. As a consequence, many workers have chosen to treat rarity as a discontinuous or categorical variable.

This approach involves determining cut-off points for abundances and range sizes, below which species are regarded as rare. The position of such cut-off points will inevitably be rather arbitrary. They could be absolute values of abundance or range size, but this would give only very limited scope for comparative studies. If we want a general definition of rarity, cut-offs will have to be relative.

There are several ways in which relative cut-offs can be applied. I will consider three.

- Proportion of species: rare species are defined as the x% with the lowest abundances or smallest range sizes in the assemblage.
- Proportion of sum: rare species are defined as those with abundances less than x% of the summed abundances of all species in the assemblage, or as those with a range size less than x% of the largest range size possible in the study area.
- Proportion of maximum: rare species are defined as those with abundances or range sizes less than x% of those of the species that have the highest abundance and the largest range size in the assemblage.

The application of these methods to a data set is illustrated in Table 1.2. The most appropriate value of x may vary for the different methods and need not

4   What is rarity?

Table 1.2 Examples of the delineation of rare species (bracketed) using proportion of species (Spp), proportion of sum (Sum) and proportion of maximum (Max) definitions, with cut-off points of (a) 25%, and (b) 5%. Data are the mean numbers of dung beetles of different species caught per trap in the spring at a site in the Mediterranean, from Lumaret and Kirk (1991)

	Species	Abundance	Spp	Sum	Max
(a)	1	1			
	2	1			
	3	1			
	4	2			
	5	3			
	6	5			
	7	5			
	8	7			
	9	10			
	10	13			
	11	18			
	12	21			
	13	28			
	14	31			
	15	49			
	16	67			
	17	97			
	18	107			
	19	130			
	20	1685			
	Total	2281			
(b)	1	1			
	2	1			
	3	1			
	4	2			
	5	3			
	6	5			
	7	5			
	8	7			
	9	10			
	10	13			
	11	18			
	12	21			
	13	28			
	14	31			
	15	49			
	16	67			
	17	97			
	18	107			
	19	130			
	20	1685			
	Total	2281			

be the same for abundances and range sizes. If the definition of rarity is to be general, the same values would, however, have to be used with all assemblages.

These different methods of applying cut-offs have varying strengths and weaknesses. A proportion of species methods results, by definition, in at least some species in any assemblage being labelled as rare (unless the assemblage is sufficiently species poor that x% is a fraction of a species). Using either of the other methods the proportion of species that are defined as rare would depend entirely on the shapes of the frequency distributions of species' abundances and range sizes. Since neither of these are constants but may take radically different forms (Chapter 2), these proportions may vary markedly. It is quite possible with these methods that there may be assemblages for which no species fall into the rare category. Using the proportion of sum method, it is also possible that assemblages would exist in which all species would be classified as rare. In terms of the proportions of species that are categorized as rare the proportion of species method would seem the most appropriate, as it is generally accepted that assemblages typically contain some rare and some common species. It also has the probably desirable property that the number of rare species in an assemblage is a direct function of species' richness. The more speciose the assemblage, the more species will be classified as rare. How numbers of rare species relate to the richness of an assemblage using either the proportion of sum or the proportion of maximum definition is less obvious. Though they would tend to increase with increasing richness in both cases, the function by which they do so may be case specific.

A further advantage of the proportion of species definition, as against either the proportion of sum or the proportion of maximum, is that it requires less information to delineate rare species. A simple ranking of species' abundances or range sizes, or recognition of the x% of species with the smallest abundances and range sizes is sufficient. If the proportion of sum definition is to be based upon abundances it is necessary to be able to estimate the actual abundances of all the species in the assemblage to calculate their summed abundance. If the proportion of maximum definition is to be applied it is necessary to estimate the abundance or range size, respectively, of the most abundant or widespread species, to rank the other species on the appropriate variable, and to determine the abundances or range sizes of the other species in the vicinity of the cut-off point sufficiently accurately so that this point can reliably be defined.

All three of the definitions suffer from the same potential drawback because they are based on relative rather than absolute criteria. That is that, through time, a particular species may move in and out of the rare category even though it maintains a stable abundance and spatial distribution and the composition of the assemblage remains constant. This may happen because changes in the abundances or range sizes of the other species determine whether or not it falls above or below the chosen cut-off point.

Tables 1.3 and 1.4 are collations of the criteria by which rare species have been delineated in a number of studies (many additional authors provide no

Table 1.3 The criteria by which a number of studies have delineated rare species on the basis of abundance and/or range size. The objectives of the different studies were varied (including avoidance of statistical biases and determining the amount of editing of distributional data required before producing maps). Definitions of classes of lower abundance or distribution than rare, if any, are given in square brackets

Source	Taxon	Criteria
Beebe (1925)	Various	Observed but seldom. [Unique, recorded but once from the research area. Probable, present in numbers just outside the research area]
Dony (1953, 1967)	Plants	Frequencies were calculated by visual impression
Bowen (1968)	Plants	Less than a thousand plants in any locality
O'Neill and Pearson (1974)	Birds	Seen regularly but in small quantities. [Isolated, seen irregularly. Accidental, recorded not more than twice]
Perring and Walters (1962)	Plants	Recorded from ≤20 vice-counties (<14% of total)
Ridgely (1976)	Birds	Recorded on fewer (usually considerably fewer) than 25 % of trips in proper habitat and season (a 'trip' is considered to be a day's field work) [Very rare, records extremely few anywhere in Panama, and always in small numbers, but presumed to be a resident within the country, or within the expected range of a migrant or wanderer]
McGowan and Walker (1979)	Copepods	$<10^2$ individuals in the 62 samples analysed. [Very rare, <10 individuals]
Ridgely and Gaulin (1980)	Birds	Recorded only occasionally
Roberson (1980)	Birds	Has occurred, during the most recent 5 year period, an average of four times or less in any West Coast state or province
Werner (1982)	Anurans	Seen (or heard) less than 25% of the time (less than one sighting for every four trips)[a]
Hartshorn and Poveda (1983)	Trees	0.1–0.01 mature individuals/ha. [Very rare, <0.01/ha]
Stiles (1983)	Birds	Seen regularly at longer intervals [than Uncommon, defined as one or a few seen at frequent intervals, usually not daily] in small numbers. [Occasional, seen sporadically; usually at longer intervals, in small numbers (but also applies to occasional flush 'invasions'). Accidental, five or fewer records to date]
DeSante and Pyle (1986)	Birds	Generally not expected on any given day. Typically no more than a few are detected on fewer than 10 % of the days. The actual number of occurrences for a state or province in any given season can vary from as few as 11 recent records to as many as 30 records/year. [Extremely rare, ten or fewer records for the past 50 years for the state or province in the given season]

Table 1.3 continued

Source	Taxon	Criteria
Jefferson and Usher (1986)	Plants	Occurring in 50 or fewer 10 km squares in the British Isles
Bushnell *et al.* (1987)	Invertebrates	One organism collected during a field season
Youtie (1987)	Insects	Fewer than 10 individuals collected
Goodman *et al.* (1989)	Birds	1–100 individuals ('abundance designations are generally based on the authors' subjective estimate of the maximum number present in the country on any given day. This includes the number of passage and winter visitors, and for residents the number of breeding pairs')
Tonn *et al.* (1990)	Fish	Occurring in five or fewer lakes, out of possible totals of 113 and 51
Verkaar (1990)	Plants	Occurring in <30 25 km^2 grid squares, out of a total of 1677 in the Netherlands
Basset and Kitching (1991)	Arthropods	One individual collected
Buzas and Culver (1991)	Foraminifera	Recorded from 1 or 2 localities in a geographic area
Landolt (1991)	Plants	Up to 200 individuals or very local [very rare, up to 20 individuals]
Laurance (1991)	Mammals	<1% of all captures
Morgan *et al.* (1991)	Birds	Average relative abundance of 0.01–1.00
Avise (1992)	Various	<10^4 individuals

[a] Criterion suggested, but not applied to any data.

indication of the criteria by which species were designated as belonging to different categories). In virtually all instances these criteria are based on measures of species abundances or levels of spatial occurrence. Almost none of the studies state the rationales by which cut-off points were arrived at. Moreover, although rare species may, for example, have been defined as those having abundances that are less than x% of the summed abundances of all the species in an assemblage or of the largest possible range, one cannot usually tell whether this value was chosen because it is a definition the authors usually apply or because the number of rare species it generates seems to be about right! One suspects that it is often the latter. That is, what appear to be proportion of sum criteria are used to apply a proportion of species definition of rarity. Clearly, the net result of the variety of methods is that the proportion of species eventually treated as rare in a study varies enormously (Table 1.4).

Table 1.4 The criteria by which a number of studies have been delineated as rare species, and the proportion of the total number of species in the assemblage (N) which these comprise (% Spp.). In all cases the numbers refer to the spatial scale at which rarity was defined. The objectives of the different studies were varied

Source	Taxon	N	Criteria	% Spp.
Hall and Moreau (1962)	Birds Ethiopian	1700	Range does not extend more than 250 miles in any direction	5.6
	Palaearctic	1100		<1.8
	Nearctic	750		1.2
	Australia	520		2.3
Munves (1975)	Birds	81	Seen only a few times	9.9
Karr (1977)	Birds Panamanian site	172	Seen only once or, at most, a few times during the study	44.8
	Costa Rican site	331		24.8
Pearson (1977)	Birds Sites in:		<0.04 sightings per hour of observation	
	Ecuador	254 (95)[a]		79.5
	Peru	214 (83)		81.8
	Bolivia	207 (83)		76.8
	Borneo	142 (34)		59.9
	New Guinea	114 (31)		49.7
	Gabon	154 (42)		72.7
Thomas (1979)	Birds	243	Found less than five times	14.8
Haila and Järvinen (1983)	Birds	121	Observed at most five times in line transect counts	37.2
Pickard (1983)	Angiosperms	180	Common and very uncommon species occurring on <10% of 437 m grid squares and abundant species on <5% of squares	27.8
	Ferns	43		27.9
Maitland (1985)	Fish	54	Native species with only a few known populations in the British Isles, or native species the populations of which are potentially unique individually and decreasing in number	14.8[b]
Goeden and Ricker (1986)	Insects on *Cirsium californicum*	58	Collected at one or two sites from a possible 28	50.0
	Cirsium proteanum	31	Collected at one or two sites from a possible 12	71.0
Hubbell and Foster (1986)	Plants	303	Average density <1 individual/ha	36.6
Usher (1986a)	Plants	65	Not more than 10 individuals or three clumps in the field	26.2

Table 1.4 continued

Source	Taxon	N	Criteria	% Spp.
Roubik and Ackerman (1987)	Orchid-bees			
	Site SR	46	log. of mean annual abundance <0.6	23.9
	Site PR	43		27.9
	Site CC	50		26.0
C. Nilsson *et al.* (1988)	Plants	366	Recorded from one to two sites out of a possible 149[c]	23.5
Adsersen (1989)	Plants	604[d]	Rarely collected, or occur only in areas where human exogenous impacts threaten the vegetation structure or the species itself	23.8[d]
Deshaye and Morisset (1989)	Plants	271	Occurred in 1–5 island-habitats (sum of the numbers of habitats on every island in which this species was present) out of a possible 248	31.4
Dzwonko and Loster (1989)	Plants	114	Occurred in ≤1/10 of localities	49.1
Faith and Norris (1989)	Macro-invertebrates	269	Have abundances comprising ≤0.5% of total abundance of all taxa	92.6
Rands and Myers (1990)	Amphibians	52	Rarely seen[e]	13.5
	Reptiles	82	Rarely seen	18.3
Longton (1992)	Mosses	692	Recorded in 15 or fewer 10 km squares of the British national grid since 1950 during an extensive field survey	25.6[f]
Osborne and Tigar (1992a)	Birds	285	1–100 individuals	49.5

[a] Figures in parentheses are the number of unseen but potentially occurring species included in the rare class.
[b] 19% using total for native species only (12 are introduced).
[c] In this instance the stated criterion was actually the proportion of species desired to be classified as rare, and the numbers of sites was the cut-off point to achieve this.
[d] Species + subspecies.
[e] Includes species which are possibly extinct.
[f] Includes about 15 species that appear to be extinct in Britain.

1.3   SCALE DEPENDENCE AND INDEPENDENCE

Whatever definition of rarity one uses, the results it gives will be influenced by the spatial scale at which it is applied. The studies in Tables 1.3 and 1.4 define rarity at a variety of spatial scales. Thus, Hall and Moreau (1962) define it at the level of entire biogeographic regions, and Karr (1977) defines it on the basis of occurrences on his study plots. While some would clearly prefer to reserve the term 'rare' for use only with respect to species' global populations and entire geographic ranges most would accept that there will be some species, albeit different ones in each case, that can be regarded as rare at any given spatial scale (e.g. Harper, 1981). A consequence of the latter position is that a species may be rare on one scale but not on another. Pigott (1981), for example, argues that while *Tilia platyphyllos* might be regarded as the rarest native species of tree in northern Europe, it is widely distributed in Europe as a whole and would not be regarded as rare on this scale (Figure 1.1).

While it is generally accepted that 'abundance' may be determined at virtually any spatial scale, it is something of a departure from what is accepted as normal usage to use 'range' equally broadly (although a few authors have applied the term at several scales). In particular, 'range' is not often applied to spatial occurrence at a local scale. To do so, does, however, have the great advantage of minimizing confusion between the various meanings of the alternative word 'distribution'. Because much of this book will be concerned both with the spatial 'distribution' of individuals and with the form of frequency 'distributions', I have chosen to avoid this potential confusion as

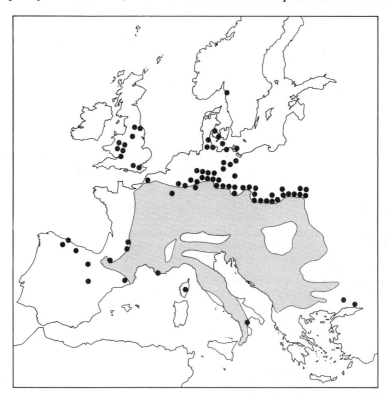

Figure 1.1 Occurrence of the tree *Tilia platyphyllos* in Europe. (Redrawn from Pigott, 1981.)

much as possible by regarding 'range' as a generic term covering all spatial scales. For greater precision, where necessary, three different prefixes will be used. 'Micro-range' will be used to mean a distribution at a local scale (a small area of homogenous habitat), 'meso-range' to mean a distribution at a regional scale (an area large enough to embrace many habitats, but not so large as to encompass the entire geographic ranges of a significant proportion of the species in an assemblage), and 'macro-range' will be used interchangeably with 'geographic range' to mean a distribution at a biogeographic scale (an area large enough to encompass the entire geographic ranges of the majority of the species in an assemblage). Of these, the area over which meso-ranges are measured is perhaps the most liable to be arbitrary. Typical limits might be those of a large true island, which would be reasonably natural, or the boundaries of a mainland state or country, which frequently would not be so.

The particular spatial scale at which a study is performed may not only determine which species are regarded as rare. It may also have profound effects on which processes have given rise to that rarity and to any patterns that are observed in it. This will be a repeated theme of later deliberations (especially in Chapters 4 and 6).

Although the categorization of species as rare or otherwise (or, using a continuous definition, as rare and less rare) is made with respect to one particular spatial scale, this does not mean that this is the only scale at which this categorization may be of interest. One might, for example, determine which species are and which are not rare on the basis of their global abundances or range sizes, and then explore how the numbers of these rare species that are present differ between regional or local assemblages of varying sizes or of the same sizes in different places. Thus, White *et al.* (1984) examined how the numbers of rare species, defined on the basis of their restricted range at state and national levels, changed between much smaller areas that differed in their overall species richness.

The recognition that rarity can be defined at different spatial scales has some important consequences. Foremost, it suggests a way out of the apparent contradictions in widespread beliefs that, for example, tropical organisms tend to be rare, and that within the tropics some species are common. Both statements can be true, dependent upon the scales at which rarity is determined. Tropical organisms may tend to be rare at a global scale, while at the scale of the tropics many will not be regarded as such.

Of course, although the concept of rarity can be applied at almost any spatial scale, it is of primary interest and has been most extensively studied at regional or biogeographic levels. Discussion of rarity in this book will reflect that emphasis.

1.4 ENDEMISM AND RARITY

At large scales, the concept of rarity is closely allied to that of endemism. Species are endemic to an area if they occur within it and nowhere else. They will thus tend to have smaller range sizes and abundances than those species which are not endemic. Endemism and rarity are not, however, interchangeable. Species may be endemic to an area and yet occur throughout it at levels of abundance or occurrence greater than those of many, or even most, of the other species found there. Likewise, at the level of some islands, virtually all the species are endemic, yet it would be of little value to regard them all as rare at this scale. Kruckeberg and Rabinowitz (1985) state, 'The narrow or local endemic is the one that best fits the colloquial notion of rarity. However, the term endemism, in its classical biogeographic usage does not necessarily imply rarity or even small range'.

1.5   DELINEATING AN ASSEMBLAGE

Just as the concept of rarity can be applied at different spatial scales, it can also be applied to assemblages the bounds of which are constrained in different ways.

1.5.1 Taxonomic constraints

Perhaps the most obvious limits that can be manipulated are taxonomic. Increasing or decreasing the numbers of taxa which are included as members of an assemblage will change which species are categorized as rare. There may be two reasons. First, even though the methodology used remains constant, the numbers of species so defined are liable to change. Second, while most taxa have species which have very few individuals and very small range sizes, they may vary drastically in the maximal abundances and range sizes which species attain.

Again, as with spatial scales, species can be categorized as rare with reference to a broad assemblage, and then the classifications used to address questions about the distribution of rare species among their component taxa. Hall and Moreau (1962) find that, by their definition, 5.6% of birds in the Ethiopian region are rare (Table 1.4), but that these comprise 8% of the passerines, and within the passerines 6% of the Estrildidae, 5% of the Muscicapinae, 12% of the Nectariniidae, 11% of the Ploceidae, 13% of the Sylviinae and 14% of the Turdinae.

1.5.2 Vagrancy

Regardless of the taxonomic limits placed on an assemblage, not all of the species which it contains have equal status. In particular, at local and regional scales, many are not permanent members of the assemblage, do not breed, or do not have self-sustaining populations in the area of interest. Such species have been variously termed accidentals, casuals, immigrants, incidentals, strays, tourists, transients and vagrants; I shall refer to them as vagrants. Typically, though not necessarily, they occur at low abundances and have small

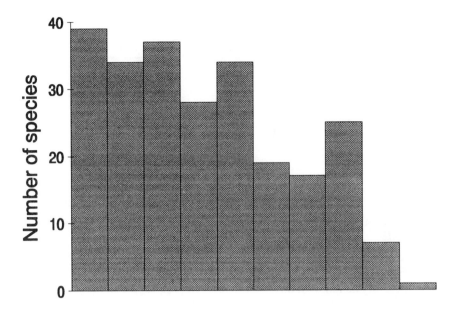

Figure 1.2 Frequency distribution of the numbers of accepted records of rare bird species in Britain in the period 1958–1989. Abundance classes are 1 individual, 2–3, 4–7, 8–15 etc. (From data in Whiteman and Millington, 1991.)

range sizes in a study area. If included in classifications, they stand a disproportionately high probability of being categorized as rare. The vast majority of the bird species regarded as rare in Britain are vagrants and have very low abundances (Figure 1.2).

The criteria by which vagrants are recognized are varied and, as in delineating species which are rare, cut-off points are ultimately arbitrary. Notwithstanding, they are generally regarded as constituting an often high proportion of species richness. Thus, Hubbell and Foster (1986) regard many of the rare species of trees occurring on their 50 ha study plot of old-growth forest on Barro Colorado island as immigrants established from second-growth forest. Osborne and Tigar (1992a) record large numbers of the less abundant birds of Lesotho as non-resident (Table 1.5), with among the probable non-breeding visitors 49 species visiting the country regularly and a further 51 only occasionally. Pimentel and Wheeler (1973) class almost half of the arthropods encountered in their study of alfalfa as incidental to the system (Table 1.6).

Shmida and Wilson (1985) argue that mass effects, the flow of individuals of species into areas where they cannot maintain viable populations (i.e. vagrants), are one of four biological determinants of species richness (the others being niche relations, habitat diversity and ecological equivalency). Following many others, Stevens (1989) proposes that mass effects are

Table 1.5 Numbers of species of different status in the lower abundance categories of the birds of Lesotho. Figures are for confirmed and suspected breeding, those in parentheses for confirmed breeding only. Very rare, 1–10 individuals; rare, 10–100; scarce, 100–1000; and special, species which although not scarce in Lesotho are of regional or international importance. (From Osborne and Tigar, 1992a.)

| | Resident | Non-resident | |
	Breeding	Breeding	Non-breeding
Very rare	12 (5)	5 (2)	73
Rare	27 (7)	2 (0)	22
Scarce	25 (18)	5 (5)	5
Special	5 (5)	1 (1)	0
Totals	69 (35)	13 (8)	100

especially important in generating the high levels of local richness observed in much of the tropics.

How vagrants are best treated in the context of rarity will very much depend upon the questions that are being asked. From a conservation perspective they should usually be ignored, on the grounds that because they cannot maintain viable populations they are not suitable candidates for management, or at least there are almost certainly areas where management would be far more effective. From an ecological perspective, however, vagrants may perhaps best be regarded as part of the assemblage because they contribute to a potentially large number of between-species' interactions.

Table 1.6 Number of arthropod species of different status in an alfalfa community. Herbivores: primary, development completed on alfalfa; secondary, some nourishment derived from alfalfa; incidental, no feeding observed to occur on alfalfa. Predators (and Parasites): primary, development completed while feeding on prey in the alfalfa community; secondary, any of its states observed to take prey from alfalfa; incidental, no prey was observed taken from alfalfa. (From Pimentel and Wheeler, 1973.)

	Number of spp.		Number of spp. (%)	
Primary herbivore	46	All primary	138	(23.4)
Secondary herbivore	145	All secondary	196	(33.2)
Incidental herbivore	121	All incidental	257	(43.5)
Primary predator	46			
Secondary predator	51	Total	591	
Incidental predator	119			
Primary parasite	46			
Incidental parasite	17			
Total	591			

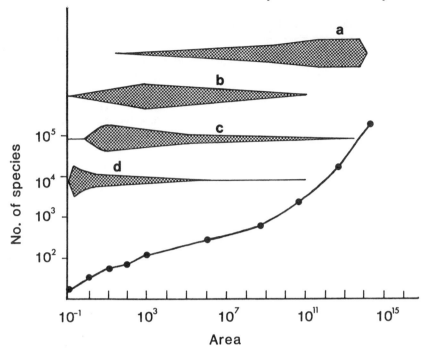

Figure 1.3 Nested species–area relationship for plants, spanning samples in the mattorral in Israel to the entire world (area measure in m^2). The hatched areas represent the hypothesized relative contribution of biological determinants to species' richness at different spatial scales; a, ecological equivalency; b, mass effect; c, habitat diversity; and d, niche relations. (Redrawn from Shmida and Wilson, 1985.)

The magnitude of the complications which vagrants generate may be scale dependent. Shmida and Wilson (1985) propose that mass effects are most important at meso-scales (Figure 1.3). Their effects will decline at scales which exceed the majority of dispersal distances of species, while at very small scales the numbers of vagrants are limited by the size of the target area. Nonetheless, vagrants are almost invariably present at any scale less than the global.

1.6   RARITY AND VULNERABILITY

Thus far, I have not mentioned what is perhaps the most widespread formal use of the term 'rare' in biology. That is, in schemes of classification which have been developed to categorize species on the basis of their supposed risk of extinction. There are many of these schemes. Perhaps the best known is that presently used by the International Union for Conservation of Nature and Natural Resources (IUCN), in which species are variously categorized as extinct, endangered, vulnerable, rare, indeterminate, out of danger, or insufficiently known (Table 1.7). This categorization, and its variants, forms the basis

Table 1.7 Definitions of the IUCN Red Data categories. (From Davis *et al.*, 1986.)

Extinct (Ex)
Taxa which are no longer known to exist in the wild after repeated searches of their type localities and other known or likely places.

Endangered (E)
Taxa in danger of extinction and whose survival is unlikely if the causal factors continue operating. Included are taxa, whose numbers have been reduced to a critical level or whose habitats have been so drastically reduced that they are deemed to be in immediate danger of extinction.

Vulnerable (V)
Taxa believed likely to move into the Endangered category in the near future if the causal factors continue operating. Included are taxa of which most or all of the populations are decreasing because of over-exploitation, extensive destruction of habitat or other environmental disturbance; taxa with populations that have been seriously depleted and whose ultimate security has not yet been assured; and taxa with populations which are still abundant but are under threat from serious adverse factors throughout their range.

Rare (R)
Taxa with small world populations that are not at present Endangered or Vulnerable, but are at risk. These taxa are usually localized within restricted geographical areas or habitats or are thinly scattered over a more extensive range.

Indeterminate (I)
Taxa known to be Extinct, Endangered, Vulnerable or Rare but where there is not enough information to say which of the four categories is appropriate.

Out of Danger (O)
Taxa formerly included in one of the above categories, but which are now considered relatively secure because effective conservation measures have been taken or the previous threat to their survival has been removed. In practice, Endangered and Vulnerable categories may include, temporarily, taxa whose populations are beginning to recover as a result of remedial action, but whose recovery is insufficient to justify their transfer to another category.

Insufficiently Known (K)
Taxa that are suspected but not definitely known to belong to any of the above categories, because of lack of information.

of the *Red Data Books* (e.g. Williams and Given, 1981; Perring and Farrell, 1983; see P. Scott *et al.*, 1987 for a historical review). Such lists may serve several functions, for example: as a resource document for conservation (e.g. planning reserves, controlling or modifying development, formulating management plans); providing a focus for the centralized input of new information; as a useful information source for publicizing the plight of endangered species; for legislative purposes; and as a focus for planning research (McIntyre, 1992).

Munton (1987) provides an excellent review of the methods by which species have been categorized as under different degrees of threat. In the course of this review, he placed a selection of the categories that have been used into different classes. By and large, those schemes which included a rare category

Table 1.8 A selection of 57 categories from schemes classifying species on the basis of threat, placed in various classes. Categories in brackets occur in more than one class. (From Munton, 1987.)

Species has disappeared
Extinct, Probably extinct, Extirpated recently, Extirpated, Known only from osseous remains, Species presumed extinct, Practical extermination, Absolute extermination, Extermination in wild state

Species is under threat
Threatened, Threatened phenomenon, Commercially threatened, Community endangered through trade, Almost extinct, Species is likely to become endangered, Very gravely threatened, Potentially threatened, Threatened with early extermination, Species in great danger, Threatened with extinction

Species is declining
Declining species, Endemic and slowly decreasing, Depleted, (Very rare and decreasing)

Species is found only in small numbers
Rare, Very rare and decreasing, Extremely rare, Very rare but believed to be stable or increasing, Less rare but decreasing, Formerly rare but no longer in danger, Some rare birds probably not in immediate danger, Small populations, Less rare but believed to be threatened, Unique, The rarest, Exceptionally rare, Sufficiently rare

Species is only found in a small area
Peripheral, Disjunct, Limit of range relict, Endemic (Endemic and slowly decreasing), Restricted local

There is a lack of data on species status
Undetermined, Hypothetical, Insufficiently known, Status undetermined, Status inadequately known – survey required, data sought

Miscellaneous
Care demanding, Additional species, Out of danger, Neither rare nor threatened, Migratory

Monitoring needed
Situation à surveiller, Species need monitoring

---

of some description defined it as containing those species which were only found in small numbers (Table 1.8). Some more detailed examples are provided in Table 1.9. Munton states, however, that 'rare' is the most confusing of all the categories which are commonly used, because opinion is divided over the significance of the phenomenon. He also recognizes two problems associated with the 'rare' category. First, by ignoring differences in the spatial distribution of small populations it confounds species which may need to be protected in different ways (e.g. a species with a small population located at a single site and a species with small population that is spread over a large region). Second, because by itself rarity does not indicate a species' risk of extinction (see Chapter 7) it would be more logical to use rarity as a parameter for the assignment of species to different categories of threat rather than as a category itself. Indeed, such an approach has essentially been taken in proposals to redefine the IUCN categories of threat (Mace and Lande, 1991).

Table 1.9 Examples of definitions of rarity from schemes classifying species according to how threatened they are. From the collation of Munton (1987), in which the full references can be found

---

● Ashton, R.E. (1976) Endangered and Threatened Amphibians and Reptiles of the United States
  Rare: Those species that are considered rare throughout the state or are found in environmental conditions disjunct from the normal geographic range of the species.

● Ayensu, E.S. and De Philipps, R.A. (1978) Endangered and Threatened Plants of the United States
  A *rare* species of plant is described as one that has a small population in its range. It may be found in a restricted geographic region or it may occur sparsely over a wide area.

● Frugis, S. and Schenk, H. (1981) Red List of Italian Birds
  Rare species: Species present in Italy with small populations which at present are not threatened nor considered vulnerable but whose 'natural' rarity puts them in peril. Status:
    Species which in Italy are on the edge of their geographical range.
    Species whose populations are very local within their range or which are present with very low density even on a wider range
    Species of recent (post 1950) establishment in Italy and whose populations need special conservation measures to facilitate their spreading into suitable habitats and their permanent establishment

● Given, D.R. (1981) Rare and Endangered Plants of New Zealand
  Rare (R): Only small populations are known or the species is found only in restricted areas where it may be locally common. For the most part however the numbers of plants and localities where it is found are reasonably stable.

● Heintzelman, D.S. (1971) Rare and Endangered Fish and Wildlife of New Jersey
  Rare: A rare species is not presently threatened with extinction, but it occurs in such small numbers in New Jersey that it may become endangered if its environment deteriorates further or other limiting factors change. Careful watch of its situation is essential.

● Miller, R.R. (1972) Threatened Freshwater Fishes of the US
  Rare: Not under immediate threat of extinction, but occurring in such small numbers and/or in such a restricted or specialized habitat that it could quickly disappear.

● Tanasiychuk, V.N. (1981) Data for the 'Red Book' of Insects of the USSR
  Rare species, not yet directly threatened with extinction, but occurring in small numbers or in such small areas they may rapidly disappear.

The variety of ways in which the 'rare' category has been constrained in the generation of lists of threatened species and in population and community biology more broadly, makes for complications in drawing parallels between the two sets of schemes. Nonetheless, some rough-and-ready rules can be suggested. In most instances, species which are listed as rare in terms of their risk of extinction, together with those species which are regarded as being under threat, will also be viewed as rare in strictly ecological studies when rarity is defined at the same spatial scale (although there will be exceptions). The converse is less likely to be true there are frequently many species which are regarded as rare in ecological categorization schemes, but do not enter lists of threatened species.

1.7   THE USE OF 'RARITY' IN THIS BOOK

Drawing together various considerations (e.g. ease of use, past usage, versatility), I favour a discontinuous definition of rarity based on a proportion of species' constraint. More particularly, I suggest that a useful cut-off point is the first quartile of the frequency distribution of species abundances or range sizes (i.e. a cut-off of 25%). From a practical perspective 25% is a convenient figure, as it is frequently possible to determine this group of species. With poor sampling, many species may be recorded as having low abundances or small range sizes of the same magnitude, thus it is more difficult to determine those species comprising smaller proportions. A cut-off of 25% is also not greatly at odds with many of the studies listed in Table 1.4. By way of shorthand, this will be referred to as the quartile definition.

Where convenient, the quartile definition has been applied throughout this book. It has proved a useful tool in formulating many of the ideas expressed. Its application has been rather more limited than one might have hoped, because it is often difficult to apply retrospectively to published studies. The same would, however, be true of virtually any other strict definition. Perhaps one of the greatest advances to the study of rarity would be the establishment of recognized criteria for its identification, or at least the analysis of data using such criteria in parallel with any other criteria thought by individual workers to be preferable. Those species defined as rare under the quartile definition are indicated in many of the figures used in subsequent chapters; they are delineated on abundance and range size on both axes, by dashed lines.

Some additional points about the use of rarity in this book should also be made. First, in the spirit of minimizing the prejudgment of the causes and consequences, rarity is defined in terms of abundance *or* range size. The interactions between the two classifications, which result for the same assemblage, are explored at some length (Chapter 3). The temptation to regard species falling in or out of the first quartiles of the frequency distributions of species' abundances and/or range sizes as forming different kinds of rarity or non-rarity has been resisted. There have been many attempts to recognize, and sometimes label, different forms of rarity (Griggs, 1940; Good, 1948; Mayr, 1963; Drury, 1974; Stebbins, 1978a; Terborgh and Winter, 1980; Rabinowitz, 1981a; Main, 1984; Cody, 1986; Soulé, 1986; Rabinowitz *et al.*, 1986; Arita *et al.*, 1990; Bawa and Ashton, 1991; McIntyre, 1992). Of these, the most well-known is that of Rabinowitz (1981a; Rabinowitz *et al.*, 1986) who categorized plant species according to geographic range (large or small), local population size (large or small) and habitat specificity (wide or narrow). Of the eight possible combinations of these states (Table 1.10) she recognized seven as constituting different forms of rarity (the eighth group (common species) have large populations and ranges, and wide specificities). In a similar

Table 1.10 The typology of rare species proposed by Rabinowitz (1981a). With the exception of the first which is listed, all combinations of the two states of each of the three characteristics are regarded as forms of rarity. Plant examples are given for various of the combinations (Rabinowitz, 1981a)

Geographic range	Habitat specificity	Local population size	
Large	Wide	Large, dominant somewhere	Locally abundant over a large range in several habitats (fat hen *Chenopodium album*)
Large	Wide	Small, non-dominant	Constantly sparse over a large range and in several habitats (knotroot bristle grass *Setaria geniculata*
Large	Narrow	Large, dominant somewhere	Locally abundant over a large range in a specific habitat (red mangrove *Rhizophora mangle*)
Large	Narrow	Small, non-dominant	Constantly sparse in a specific habitat but over a large range (*Taxus canadensis*)
Small	Wide	Large, dominant somewhere	Locally abundant in several habitats but restricted geographically (pygmy cypress *Cupressus pygmaea*)
Small	Wide	Small, non-dominant	Constantly sparse and geographically restricted in several habitats (Non-existent?)
Small	Narrow	Large, dominant somewhere	Locally abundant in a specific habitat but restricted geographically (*Shortia galacifolia*)
Small	Narrow	Small, non-dominant	Constantly sparse and geographically restricted in a specific habitat (*Torreya taxifolia*)

vein, Bawa and Ashton (1991) distinguish four kinds of rarity in tropical forest trees, species that are uniformly rare, species that are common in some places but rare in between, species that are local endemics, and species that are clumped even when overall population density is very low. In the main, such approaches have necessitated arbitrary, and often subjective, divisions to the breadth of values of yet more variables (e.g. habitat specificity, dispersal ability) in addition to abundance and/or range size. In this volume these additional variables will largely be treated as continuous and their interactions with abundance and range size will be explored.

Second, in using a definition of rarity that will identify different species as rare, dependent upon spatial scale and how an assemblage is delineated, rarity cannot be regarded as a species-specific characteristic. The only way in which this might be held to be true is at the global scale for a defined assemblage.

This minimizes conflict with the use of rarity with respect to levels of threat – species may be at a high risk of extinction at some spatial scales but not at others. Just how constant in composition the rare component of a defined assemblage is with respect to space and time is explored in Chapters 4 and 5, respectively.

Third, the terms 'common' and 'widespread' will be used as antitheses of rare, when rarity is defined in terms of abundance and range size, respectively. Their application avoids repeated reference to the not rare.

2 Abundances and range sizes: measuring rarity

I attempted to carry out the figures, which seem to behave according to some mathematical formula; but when I came to deal with 3/5 of an occurrence I decided it was profitless to go on!

J. Grinnell (1922)

The concept of total areal geographic range, as one of the indissoluble characteristics of the taxonomic units species and subspecies, seems to be by no means adequately defined or understood in current zoological literature.

K.P. Schmidt (1950)

The definitions of rarity outlined in the previous chapter are based on the relative magnitudes of species' abundances or of their range sizes. There is, however, no such thing as *the* abundance or *the* range size of a species, even at a particular spatial and temporal scale. Rather, they are both generic terms, embracing several important distinctions. Answers depend on which groups of individuals are and are not included, and the methods by which abundance and range size are assessed.

In this chapter, abundance and range size are considered in more detail. The first two sections, respectively, address measures of abundance and the form of species–abundance distributions, while sections three and four address the equivalent topics with regard to range sizes. Different ways of measuring each variable and some of the associated problems are discussed, particularly as they pertain to rarity. The fifth section explores the topics of pseudo-rarity and non-apparent rarity, and the sixth the importance of phylogenetic considerations when interpreting differences in abundances and range sizes and hence rarity.

2.1 MEASURES OF ABUNDANCE

2.1.1 Different populations

Suppose we wish to ask whether species A has a different abundance from species B in a study area. We may be able to answer the question in several ways, say on the basis of counts of the entire populations of the two species, of just the adult populations or of just the reproductive populations. Some of these different populations could be counted at more than one time during each species' life cycle, and one might choose between these times. The choices which are made may determine whether species A is found to have an abundance that is comparable with, higher than, or lower than that of species B.

The potential relevance of this observation to the study of rarity is obvious. The identities of the species which are documented as having relatively low abundances and hence are categorized as rare may depend on the population for which abundances are measured. Is this important? In many instances it is probably not. The subset of abundances to which a given hypothesis is relevant can be identified, and the use of inappropriate measures of abundance avoided. Some care may have to be exercised in discerning possible pitfalls. This is especially true of analyses of the causal determinants of rarity, where it is remarkably easy to frame tests around abundances of, for example, an inappropriate lifestage.

It is not always possible to measure the abundance of some kinds of populations. This is most frequently the case for the entire population summed across all lifestages. Seed banks and prolonged diapause, for example, create essentially cryptic life stages, the enumeration of which is seldom a realistic option. In these instances it may be important to understand the extent to which species' relative abundances change, when assessed for different populations. There are no simple guidelines which enable robust predictions of the likely outcomes in any given situation. Plainly, the more similar are species' life cycles, and patterns and levels of mortality, the more similar will be their rank abundances at different individual, and groups of, life stages or ages.

Rabinowitz and Rapp (1985) explicitly tested, on a small spatial scale, whether over a $4\frac{1}{2}$ year period the rank abundances of the seedlings and shoots of prairie plant species changed as a consequence of mortality. They found no significant effect. There was no correlation between the initial abundance of seedlings and their survivorship during the study. The rank abundances of species at these life stages were, however, considerably different from those of the adult plants.

2.1.2 Individuals, biomass and cover

Many of us are used to thinking of abundances in terms of numbers of individual organisms. Indeed, this is the way in which they are enumerated in the majority of studies cited in this book. Nonetheless, we need to recognize that the broad concept of the individual is hard to apply to some groups of organisms and of little practical value in many others (e.g. for some plants, individuals can only be recognized following destructive sampling). Hence, abundances have been based on a variety of alternatives, such as biomass and vegetative cover. General considerations of rarity seldom embrace this complication, although there seems little justification for restricting its application solely to situations in which individuals can both be recognized and counted. Similarity in the composition of the species recognized as rare on the basis of numbers of individuals and of other measures of abundance is likely to be

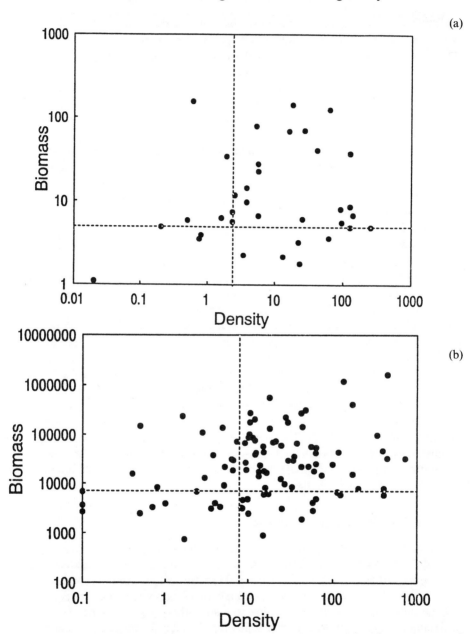

Figure 2.1 Relationships between estimates of the biomasses (kg/km^2 in (a) and g/km^2 in (b)) and densities (individuals/km^2) of mammal species, for (a) a study area in Venezuela and (b) the Neotropics. Biomasses are measured indirectly, as the product of density and the mean weight of an individual (the confidence intervals around each figure are therefore likely to be large). Dashed lines delineate those species on each axis which under a quartile definition are categorized as rare. (From data in Eisenberg *et al.*, 1979 and Arita *et al.*, 1990.)

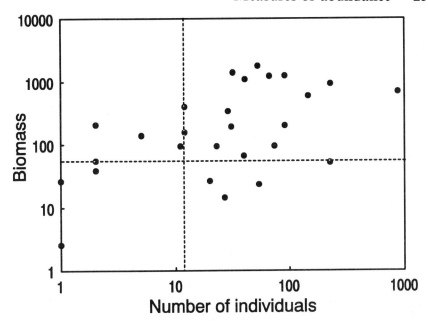

Figure 2.2 Relationship between the biomass (g) and the numbers of individuals of different fish species collected from stream sites in South Carolina in 1985. Biomasses were determined by direct measurement. Dashed lines delineate those species on each axis which under a quartile definition are categorized as rare. (From data in Meffe and Sheldon, 1990.)

variable. Figures 2.1 and 2.2 illustrate examples in which, on the basis of the estimates given, the majority of species are not rare according both to a quartile definition of rarity on the basis of numbers of individuals and such a definition on the basis of biomass. Digby and Kempton (1987) provide an example of a mollusc assemblage in which ranks of abundance and biomass differ substantially, however, in this instance the three species which are rare under a quartile definition are the same in both cases. Species of very large physical size have been identified as creating particular problems for the constancy of the ranks of species abundances and biomasses, as they are apt to rank high in terms of biomass, but low in terms of numbers of individuals. Harper (1981) makes this point with regard to bracken (*Pteridium aquilinum*), which is claimed to be one of the five most abundant plants on earth, but whose abundance in most places results from the vegetative extent and biomass of a few genetic individuals.

2.1.3 Population size and population density

In terms of numbers of individuals, abundances of a given species can be of three different kinds: the numbers of individuals in a population (population size); the number of individuals per unit area (absolute density); or the number of individuals relative to another population (i.e. some index, such as the number of individuals trapped, the number of singing males or food plant density; relative density). Estimates of relative density are usually more readily obtained than estimates of absolute density or population size, and estimates of absolute density are usually easier to obtain than estimates of population size. Many, if not most, questions in population and community biology can be phrased in such a way as to enable them to be answered using measures of relative density (Caughley, 1977). The study of rarity is one field in which it is, however, at times necessary to use all of these measures.

Perhaps the most crucial distinction is that between measures of population size and of population density. Species with similar densities in an area may have very different population sizes, while species with very different densities there may actually have very similar population sizes. This is because measures of population density, particularly locally, do not account for differences in the areal extent of populations, which may be radically different. It is on these grounds that Kubitzki (1977) has challenged the general assertion that tropical species are rare on the global scale. While they occur at low densities, tropical species may be widely distributed (a fact which tends to be ignored; Gaston, 1991a) and hence have large global population sizes. A disproportionate number of tropical species are nonetheless probably globally rare, because not only are their densities lower, but there also seems to be a decline in the mean geographic range size of species with decreasing latitude (Rapoport's rule; Stevens, 1989, 1992a).

As defined here, those species which are rare can potentially be identified using relative or absolute densities, or population sizes. Relative densities should only be used, however, where not only do intraspecific relationships between relative and absolute density exist, but where these relationships do not generate changes in inter-specific relative abundances that do not reflect changes in inter-specific absolute abundances.

Issues of local resource use and the abundance structure of geographic ranges may, for example, best be addressed using densities, while many conservation issues necessitate estimates of population sizes.

2.1.4 Rarity and abundance estimates

There are plenty of text books concerned with the methodology of generating estimates of species' population densities or population sizes and their associated sampling errors (Caughley, 1977; Southwood, 1978; Ralph and Scott, 1981; Seber, 1982; Kershaw and Looney, 1985; Taylor et al., 1985; Verner, 1985; Krebs, 1989; Goldsmith, 1991a; Bibby et al., 1992a). I do not intend to cover the same ground in any detail. In the context of rarity, there are, nonetheless, some points that should be emphasized.

(i)   Because they are scarce, a great deal of effort may be required to quantify the abundances of those species in an assemblage that are rare. Their abundances have often to be inferred or given some maximal value, because large enough samples cannot be obtained to provide more accurate estimates. Pianka (1986) observes that for some of the less abundant species of Australian desert lizards, a hundred or more man-days must be spent in the field before encountering even a single individual.

Techniques have been developed for estimating the abundances of various species which occur at either very low densities or have very small population sizes. The work of Ward *et al.*(1991) on northern spotted owls (*Strix occidentalis caurina*) provides one example.

(ii)   The greater the sampling errors around estimates of species' abundances, the less the reliability with which an assemblage can be categorized into rare and common species on the basis of these estimates and an arbitrary cut-off point (i.e. a discontinuous definition). Any tendency for sampling errors to increase proportionately for species with lower abundances, as is likely, will make the confident delineation of a group of species as rare yet more difficult.

(iii)   If sampling is not efficient enough, then many, or perhaps all, of the species in an assemblage that would have been recognized as rare may not even be recorded as present. The problem of distinguishing absences which are a result of poor sampling (sampling zeros) from those which are genuine (structural zeros) presents a significant difficulty not only for the study of rarity, but in many other topics in population and community biology (e.g. Gaston and McArdle, 1993; McArdle and Gaston, 1993). The large number of cases in which species have been stated to be globally extinct only to be rediscovered some time later (even for large organisms, such as trees, birds and mammals; Rabinowitz, 1981a; Diamond, 1985) testify to how serious this problem can be – the ecological equivalent of what is known in palaeontological circles as the 'Lazarus effect' (Jablonski, 1986a). Pearson (1977) estimates for study plots in the tropics the numbers of bird species which though not seen could potentially have occurred there. These account for 24–40% of the species total (actually observed + potentially occurring) for each plot.

While an exhaustive search of an area is usually impractical, the results of many studies would be rendered more interpretable if statements as to the confidence with which species can be claimed to be absent were used. These could be based on relationships between the number of sampling units taken, the rarity of the species, and the probability that it will be detected in a sample (McArdle, 1990).

(iv)   Estimates of the abundances of individual species that are regarded as rare with respect to some regional or global standard are not infrequently conservative. That is the estimates are significantly smaller than the actual number of individuals present. Examples include the white-breasted guineafowl (*Agelastes meleagrides*), which in the 1980s was judged, on the basis of available sources, to be one of the most threatened birds in continental Africa (Collar and Stuart, 1985), but was subsequently estimated to have a population of 30 000–40 000 birds in Taï National Park (Côte d'Ivoire) alone (Francis *et al.*, 1992). Likewise, while in the late 1970s there were thought to be only a few

hundred individuals of the Usambara ground robin (*Dryocichloides monta-nus*), this estimate was later revised to 28 000 birds (Collar and Stuart, 1985).
(v)   In some instances, the abundances of rare species may be substantially overestimated because of their rarity. For example, in those regions of the world with large numbers of active bird watchers, records of the numbers of individuals of rare species are substantially more likely to be kept than are records of the numbers of more abundant species. Bock and Root (1981) state, with respect to the Christmas Bird Census (CBC) of North America, that 'Observers will work hard to find at least one individual of any rare species which might occur in a count circle. The result is that some rare species appear to occupy the country in the ornithological equivalent of a monomolecular layer.' Indeed, no correlation could be found between carefully considered estimates of the population sizes of the Californian condor (*Gymnogyps californianus*) and whooping crane (*Grus americana*) and CBC data (Bock and Root, 1981).

The severity of such problems seems likely to depend on the patterns of data collection. Work on the records resulting from casual birdwatching in a county in Britain suggests that the documented population trends of scarce birds reflect what is known of population changes in these species elsewhere (Mason, 1990).

In general, the effects of an intense search for rare species probably mean that counts minimize real differences in abundance and do not create artificial differences (Bock *et al.*, 1977).
(vi)   A potential complication to the assessment of the relative abundances of species in an assemblage, and hence to the determination of which are rare, is produced by temporal turnover in the individuals present in a study area. Two populations which have approximately equal sizes at any one instant, but which in one case comprise the same individuals from one time period to the next, and in the other comprise different individuals, may not readily be equated. How profound a problem this poses will depend on the particular question being asked, and may to some extent be resolved by changing the spatial scale of a study. It is an issue which has been little considered in population and community biology.
(vii) The use of summary statistics derived from replicate estimates of a species' abundance can do much to ensure a more rigorous comparison of the abundances of species the numbers of which vary in space or in time. However, the choice of these statistics should be made carefully. Recent debate over the form of the relationship between a species' density and its body size demonstrates the point. A simple negative correlation between the two variables has been widely reported (Peters, 1983; Peters and Wassenberg, 1983; Damuth, 1987). However, these have typically been based upon compilations of data from the literature, which are liable to be biased towards maximal density measures simply because it is impractical to perform ecological studies in areas where species are scarce (Morse *et al.*, 1988). When abundance–body size relationships are analysed using data collected in a more uniform fashion, the simple

negative relationships frequently disappear (Morse *et al.*, 1988; Blackburn *et al.*, 1993) (abundance–body size relationships are explored further in Chapter 6).
(viii) The density of a species is derived by dividing a figure for the number of individuals present by a figure for the area in which they are contained. Even when determining the numbers of individuals is comparatively straightforward, determining an appropriate figure for the area is often not.

As early as Elton (1932, 1933), the simplistic distinction between area measures which generate *crude density* estimates (Elton's 'lowest density') and measures which generate *ecological density* estimates (Elton's 'economic density') was recognized. In the former instance, the area is defined somewhat arbitrarily, in the latter, the area is calculated such that it only includes habitats which the organism uses. Thus, crude densities will tend to be consistently lower than ecological densities. Schonewald-Cox and coworkers (Schonewald-Cox and Buechner, 1991; Schonewald-Cox *et al.*, 1991) observe that there is a general decline in measured densities as the size of a study area increases. A likely explanation for this pattern is simply that more unsuitable habitat is included with larger study areas.

Determining an appropriate ecological density measurement may not be easy. As Haila (1988) argues, each species may have several densities, depending on which habitats are included in the area measurement. Depending on the area measurement chosen, the resultant density can express widely different ecological phenomena or nothing biological at all. Haila has explored some of the consequences that can result from different methods of calculating densities. The consequences can be profound, however, this need not always be so. Different density measures may be strongly related (Figure 2.3).

2.2   SPECIES–ABUNDANCE DISTRIBUTIONS

It has often been observed that in samples from communities and in more complete censuses, most species are represented by a small number of individuals, while most individuals belong to a few abundant species (Figures 2.4 and 2.5). That is to say, the distribution of individuals among species (the species–abundance distribution) is strongly right-skewed on untransformed axes. Treating rarity as a continuous variable based on abundance, this means that, for any given assemblage, many more species will tend to have high rarity values than will have low ones. Indeed, it is not unusual to find statements to the effect that most species in an assemblage are rare, although these conflict sharply with the proportion that are typically actually categorized as rare for the purposes of studying the phenomenon (Table 1.4). The parameters of most of the discontinuous definitions of rarity that have been applied result in those species which are categorized as rare having, on average, abundances that are similar to those of a higher proportion of the total number of species in an assemblage than do those species categorized as common.

Several models have been proposed as descriptors of species–abundance distributions (e.g. broken-stick, geometric series, log-series, log-normal,

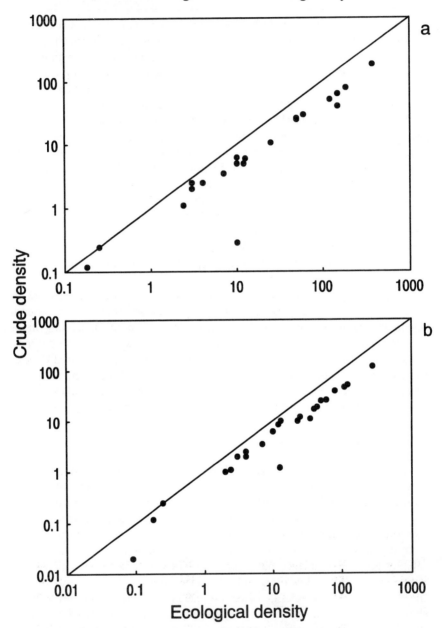

Figure 2.3 Relationships between estimates of the crude and the ecological densities (both in individuals per km^2) of non-volant mammal species in (a) the east and (b) the west side of a study area in Venezuela (different from that in Figure 2.1). Ecological densities are calculated using the area which includes only suitable habitat for the species in question. If crude and ecological densities were equal they would fall on the solid line. (From data in Eisenberg *et al.*, 1979).

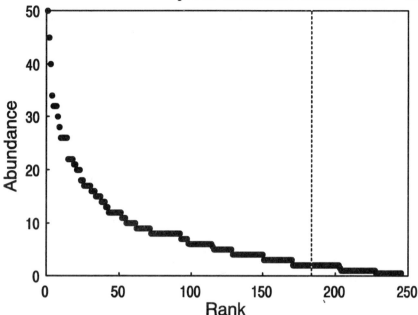

Figure 2.4 Rank–abundance (number of individuals) relationship for 245 bird species judged to be resident breeders and present at a density of 0.5 or more pairs/ha, on a 97-ha plot of floodplain plain forest in Amazonian Peru. A total of 319 species were recorded on the plot including migrants, vagrants and species with densities too low to measure. (From data in Terborgh *et al.*, 1990.)

negative binomial, Zipf–Mandelbrot). Some of these have solely a statistical basis and others have their roots in theoretical ecology. There are many reviews and critical evaluations of particular aspects, of one or more of these models (e.g. Preston, 1948, 1962, 1980; May, 1975; Sugihara, 1980; Harmsen, 1983; Hughes, 1986; Frontier, 1987; Gray, 1987; Kolasa and Strayer, 1988; Magurran, 1988; Lawton, 1990; Tokeshi, 1990). Although there would probably be general agreement over none of them, the following points can be made by way of a summary of the present situation.

- There are substantial problems in ascertaining which models best fit any given data set, and how to interpret such fits (Gray, 1987). Given that several models might fit one data set, it may be as helpful to understand why some do not.
- No single model that is presently available provides a general description of all species–abundance data. Nor is any such model likely to be derived.
- Nonetheless, some models have under some circumstances proven extraordinarily successful. Sugihara's (1980) sequential niche breakage model, in particular, has generated much interest for this reason (e.g. Kolasa and Strayer, 1988; Nee *et al.*, 1991b).

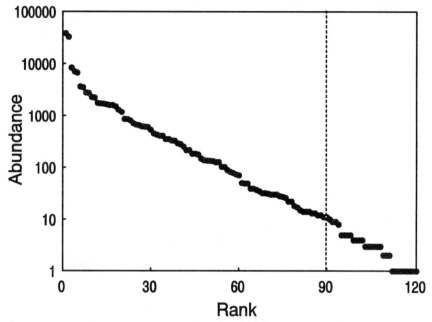

Figure 2.5 Rank–abundance (number of individuals) relationship for 120 dung beetle species caught in Mkuzi Game Reserve. Dashed line delineates those species which under a quartile definition are categorized as rare. (From data in Doube, 1991.)

- The search for ecologically motivated models has, with some notable exceptions (e.g. Hughes' (1986) dynamics model, which simulates the development and progression of a community through time) relied heavily on simple phenomenological approaches. Even where these appear to provide reasonable fits to real data it is seldom clear why, despite the huge variety of processes which affect the abundances of different species, this is so. 'The ecological theories on which the models are built often have rather tenuous links to ecological reality . . .' (Gray, 1987).
- The underlying assumptions of ecologically motivated models have not, by and large, been critically tested.
- Species–abundance distributions can be visualized as resulting from the way in which resources are divided among species and among individuals within species. Largely unresolved are the questions of how to ascertain the amount of resources available to an assemblage, and what determines how resources are divided among species and in turn are divided among individuals (see Chapter 6).

Unsurprisingly, we know least about the form of the frequency distributions of population sizes at large regional or global scales for either taxonomically or ecologically defined assemblages. The majority of studies are for substantially smaller scales. While distributions at larger scales are undoubtedly very similar to those we do know something about, knowledge of their parameters

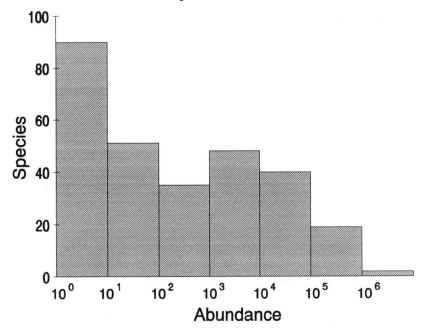

Figure 2.6 Frequency distribution of the abundances (number of individuals) of the 285 confirmed and extant bird species recorded in Lesotho since 1940. (From data in Osborne and Tigar, 1992a.)

would still represent a considerable step towards a better understanding of global patterns of rarity. If we knew how many species there were with total abundances of different sizes we could better assess the risk of loss of different proportions of these assemblages, and perhaps better prioritize how we go about conserving them.

What data there are about the abundances of species in assemblages at very large spatial scales are chiefly for bird populations at national levels (Figures 2.6 and 2.7; Merikallio, 1951, 1958; Preston, 1958, 1962; Nee *et al.*, 1991b; Osborne and Tigar, 1992a). In some cases the distributions of these data appear to conform reasonably well to a log-normal distribution. However, doubts persist over the accuracy of the sampling of species with very low abundances and the consequences these might have for the observed distributions. Recent analyses of data for breeding populations of British birds, probably the most accurately censused of any, have substantiated these concerns (Nee *et al.*, 1991b). This distribution is asymmetrical on logarithmic axes, with a left skew. That is, an excess of rare species (a tendency which has been hinted at in other studies (e.g. Preston, 1958; Haila and Järvinen, 1981; Hubbell and Foster, 1986)).

Probably the boldest attempt to explore species–abundance distributions at a very large scale has been Williams' (1960) study of the abundances of the world's insect species. This necessitated making assumptions about the probable total

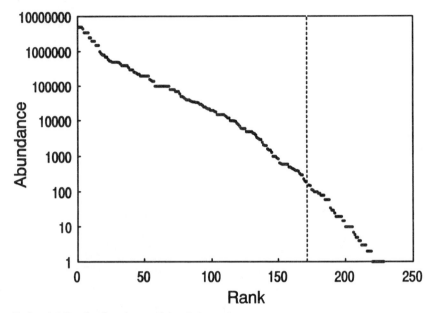

Figure 2.7 Rank–abundance plot of the estimated abundances (number of breeding pairs) of British breeding bird species. Dashed line delineates those species which under a quartile definition are categorized as rare. (From data in Marchant *et al.*, 1990.)

numbers of insect species and the sum total of individuals they comprise, and, in some instances, other parameters of the distributions. Although almost indecently speculative, this work raises important questions as to the possible numbers of species that exist at very small population sizes (a few tens of individuals or less), presumably because they are incipient or on the verge of extinction. Using very conservative estimates of the numbers of insect species, these may potentially sum to many thousands of species.

Although they tend to be the norm, at smaller scales there are exceptions to unimodal species–abundance distributions. Hanski (1991b; Hanski and Cambefort, 1991) draws attention to the frequent observation of a tendency toward approximately bimodal distributions of log-transformed abundances in dung beetle assemblages (Figure 2.8). Such distributions are uncommon in the literature at large, and it has been suggested that this may reflect the greater ease with which these assemblages can be sampled. Hanski (1991b) argues that bimodality is generated by the presence of a mixture of local and non-local species in an assemblage. Non-local species comprise the left-hand peak to the species–abundance distribution. There is good evidence that this is true for an intensively studied dung beetle assemblage in southern England (Hanski, 1991b). This finding harks back to earlier comments (Chapter 1) about the need carefully to define which species are and are not regarded as part of an assemblage before they are classified as rare or otherwise. In this instance, the species defined as rare would almost invariably include many non-breeders.

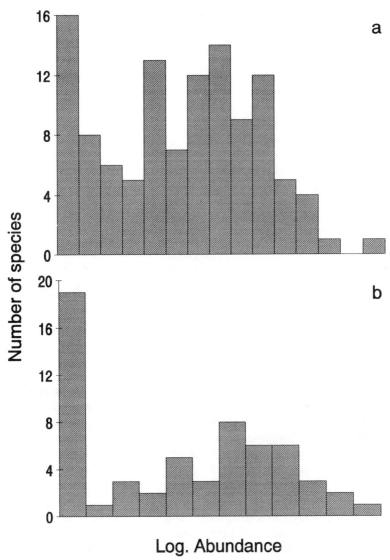

Figure 2.8 Species–abundance (number of individuals caught) distributions in two local assemblages of dung beetles, in (a) montane pasture, and (b) tropical forest. (Redrawn from Hanski and Cambefort, 1991.)

2.3   MEASURES OF RANGE SIZE

2.3.1 Which range?

Species' abundances are routinely labelled as pertaining, for example, to breeding, non-breeding, summering or wintering populations. Equivalent

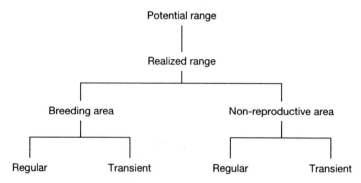

Potential range: area within which species would be found if all limitations to dispersal were overcome.
Realized range: area within which species occurs.
Regular breeding range: area within which breeding regularly occurs.
Transient breeding range: unsuitable for prolonged occupation for breeding, but species may breed here regularly for short periods.
Regular non-reproductive area: area that is periodically, but regularly, inhabited for non-breeding purposes (e.g. wintering area of migrant species).
Transient non-reproductive area: unsuitable for prolonged occupation, but species may be found here regularly for short periods for non-breeding purposes.

Figure 2.9 Scheme for the different possible components of a species' geographic range and their relationships. (Developed from Udvardy, 1969.)

kinds of tag are less often applied to species ranges. Nonetheless, several different kinds, or components (depending on one's viewpoint), of ranges can be distinguished. Figure 2.9 presents a scheme, modified from that of Udvardy (1969), for entire geographic ranges. It is essentially hierarchical. Thus, it is exceedingly unlikely that a realized range of a species is as large as its potential range, and for many species both breeding and non-reproductive areas are likely to be smaller than the realized range (e.g. Blakers *et al.*, 1984). Breeding and non-reproductive ranges will overlap to a variable extent, and the relative sizes of their regular and transient components may differ a great deal, with one or other of them being entirely absent in some instances. As with abundances, this means that the outcome of any between-species comparisons of range sizes may depend on which range is contrasted. However, within broadly similar groups of species the sizes of different kinds of ranges may be strongly correlated, as shown for the breeding and total geographic ranges of Australian birds (Ford, 1990).

The size of all of the components of the scheme presented are dynamic on some time scale. In deciding which range one is interested in, these dynamics need to be taken into account. Data on a species' spatial pattern of occurrence at large spatial scales is normally gathered over a moderate to long period, and as a result measured range sizes may not be representative of any one point in time. It seems doubtful that this will markedly affect which of any two species is regarded as having the largest range, although it could potentially do so if one of them had a range which was substantially more temporally variable in extent or position, than that of the other.

A further consideration to be made in some studies is whether one is primarily interested in present-day range sizes, or in range sizes prior to the impact of modern peoples. Commonly, historical range sizes are used rather than current range sizes (e.g. Anderson, 1977, 1984b; Pagel *et al.*, 1991), with areas from which a species has been extirpated by human activities being included and those into which it has been introduced being excluded.

2.3.2 Extent of occurrence and area of occupancy

Having determined which range one intends to measure, the decision as to how to quantify it remains. A wide variety of techniques has been used (Table 2.1). These can broadly be divided into those which measure a species' extent of occurrence, and those which measure its area of occupancy (Gaston, 1991b).

The *extent of occurrence* (which equates to the 'gross range' of some authors) expresses the distance between the spatial limits to the localities at which the species has been recorded in the region of interest (Figure 2.10). It is seldom used at scales less than a species' entire geographic range. Perhaps the simplest way to quantify a species' extent of occurrence is to sum the areas of the regions in which it has been recorded. Thus, for example, Spitzer and Lepš (1988) assign species of European noctuid moths to one of six geographic range categories, from largest to smallest, cosmopolitan, palaearctic + paleotropical, holarctic, palaearctic, eurosiberian, and european. This is a crude measure. It can be refined using more comprehensive information on the localities at which species have been recorded. Extent of occurrence is often quantified in terms of the latitudinal or longitudinal extent of these localities, or some combination thereof, or in terms of the size of the smallest area contained within an imaginary boundary line which encloses all the localities (Table 2.1). The majority of field guides and many taxonomic works depict species' ranges as a solid block of occupancy (based on marginal records), and commonly provide the basis for such calculations. Criteria (often closely related to those used in the measurement of species home ranges) are available by which the position of a boundary line can be determined, though it is usually fitted by eye.

Measures of a species' extent of occurrence include regions which, while falling within the limits of its occurrence, are not actually occupied. This may be because these areas are entirely unsuitable, or simply uncolonized at present. In calculating a species' *area of occupancy* such areas are ignored (Figure 2.10). In principle, this distinction is a straightforward one. In practice it is more complicated. There are two difficulties. First, the extent of occurrence of a species is difficult to interpret when individuals are distributed among two or more spatially isolated groups, and it certainly makes for difficult comparisons with species for which this is not so obviously the case. Methods have been

Table 2.1 Methods by which the sizes of species ranges have been measured in different studies

Meso-ranges	
North–south range	Kouki and Häyrinen (1991)
Numbers of sites at which species have been recorded	Gaston and Lawton (1988a) Gaston (1988) Collins and Glenn (1991)
Numbers of quadrats on a grid system from which species have been recorded	Lawton and Schröder (1977) Claridge and Wilson (1981, 1982) Neuvonen and Niemelä (1981) Niemelä and Neuvonen (1983) Godfray (1984) Leather (1985) Ford (1990) Gaston and Lawton (1990a,b)
Macro-ranges	
Latitudinal range	Pielov (1977, 1978) Reaka (1980) Juliano (1983) Stevens (1989) France (1992)
Longitudinal range	Reaka (1980) Juliano (1983)
Diagonal distance, calculated from north to south and east to west distances	Reaka (1980) Juliano (1983)
Maximum linear extent, measured as a straightline distance between the two most distant known localities	Juliano (1983) Kavanaugh (1985)
Area of the rectangle defined by the major perpendicular axes of the species distribution	Stevens (1986)
Area within a convex polygon	Juliano (1983)
Area within a line, usually drawn by eye, enclosing limits to occurrence	Anderson (1977, 1984a,b) Glazier (1980) McAllister *et al.* (1986) Pagel *et al.* (1991)
Numbers of geographic areas, not usually equal areas or at best only approximately so, from which species recorded	Jackson (1974) Thomas and Mallorie (1985b) McLaughlin (1992)
Numbers of quadrats on a grid system from which species have been recorded	Juliano (1983) McAllister *et al.* (1986) Schoener (1987) Ford (1990) Pomeroy and Ssekabiira (1990) Maurer *et al.* (1991)
The relative size of the biogeographic region in which species are found	Spitzer and Lepš (1988) C.D. Thomas (1991)
Numbers of sites at which species have been recorded	McAllister *et al.* (1986)

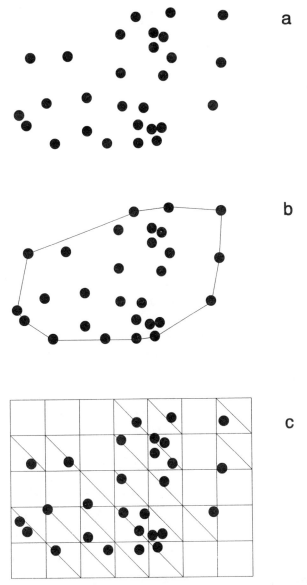

Figure 2.10 Simple model of the distinction between a species' extent of occurrence and its area of occupancy. (a) The spatial distribution of sites at which the species has been recorded; (b) one possible boundary to its extent of occurrence (the extent is measured as the area within this boundary); and (c) one measure of the species' area of occupancy, the numbers of occupied grid squares (note squares are only regarded as occupied if they contain the centre of an occupied site symbol).

Table 2.2 Forty-seven endangered species of southeast Australian plants categorized by geographic spread and recorded number of populations. (From Brown and Briggs, 1991.)

Geographic spread (km)	No. of populations									
	1	*2*	*3*	*4*	*5*	*6*	*7*	*8*	*9*	*10*
<1	22									
1–10		1								
10–50		5	1		1	1				
50–100				1	2					
100–500		1		1		2			2	
500–1000			1		1	1		1		
>1000				1						2
Total	22	7	2	3	4	4	–	1	2	2

developed to identify such isolates, with bounds often being drawn about the limits to each, their extents of occurrence calculated separately and then summed (Rapoport, 1982). However, what are identified as isolates will depend crucially upon how refined the mapping of a species' spatial distribution is. The more refined, the more isolates, and the more one tends to be measuring a species' area of occupancy and not its extent of occurrence.

The second complication is the converse of the first. The area within the bounds to a species' occurrence can be interpreted either as a reasonably refined measure of its extent of occurrence or as a crude measure of its area of occupancy. More refined measures of area of occupancy can be achieved, such as the number of squares of a grid map of the area of interest in which the species has been found, or the total number of localities at which it has been recorded (Table 2.1). The greater their refinement the more these measures will tend to depart from measures of extent of occurrence. Unlike extents of occurrence, measures of area of occupancy tend to be applied at all spatial scales.

The relationship between a species' extent of occurrence and its area of occupancy is akin to that between a crude and an ecological density. In the same way that the area measurement for density calculations should be determined by the kind of question that is being addressed, so should the choice between measures of extents of occurrence or areas of occupancy, and choice of the refinement of the area determined as the area of occupancy. The various measures of extents of occurrence and areas of occupancy will, for any given assemblage, tend to be broadly related (Table 2.2; Figure 2.11; McAllister *et al.*, 1986). Indeed, it may often be wrong to assume that the correct null hypothesis is that there is no relationship between area of occupancy and extent of occurrence. A simple quadrat-based model in which area of occupancy is increased for the minimum growth in extent of occurrence,

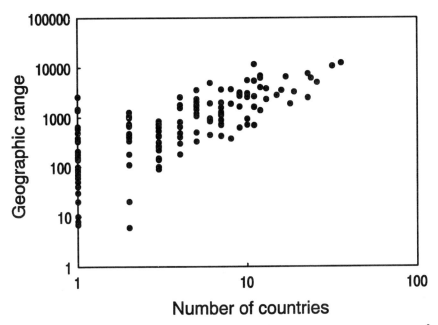

Figure 2.11 Relationship between the area within the geographic limits ($\times$ 1000 km^2) of primate species and the numbers of countries in which they have been recorded. (From data in Wolfheim, 1983.)

demonstrates a strong positive relationship between the two (Figure 2.12). Whether the rank order of species on the basis of each measure is the same will largely be case dependent.

The crudity with which range sizes are measured may have important consequences for observed relationships. Claridge and Evans (1990), for example, demonstrate that whether a significant interaction is observed between the number of insect species associated with a particular species of plant and the range of that plant may depend crucially upon the quality of the distributional data. It seems probable, however, that such examples are exceptional.

Because areas of occupancy are usually measured very crudely, the distinction between areas of occupancy and extents of occurrence may be minimal. However, the distinction can in some circumstances be an important one. Because many environmental parameters show 'reddened spectra' (Williamson, 1987 and references therein), their heterogeneity increasing with area, species' extents of occurrence will inevitably tend to be correlated with the heterogeneity in these parameters. Thus, for example, the numbers of habitats, the range of temperatures, and the range of levels of precipitation recorded within the extent of occurrence will all tend to be positively correlated with the magnitude of this extent. This may, nonetheless, say very little about the biologies of the species concerned; they may all occur within just one habitat

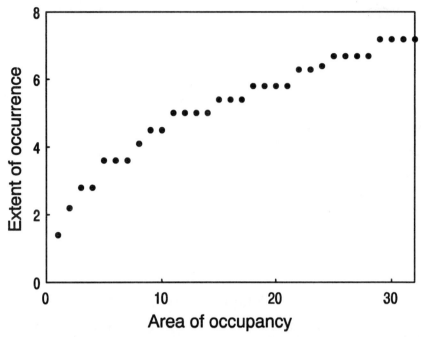

Figure 2.12 Relationship between extent of occurrence and area of occupancy for a simple model in which occupied quadrats were added in a sequence generating the minimum extent of occurrence for a given area of occupancy. Extent of occurrence expressed as units and area of occupancy as square units.

type at a narrow range of temperatures and levels of precipitation. It is only through measuring species geographic ranges as areas of occupancy that artefactual correlations with environmental factors will be avoided (see Chapter 6 for some related discussion of this problem). In fact, for most purposes, area of occupancy measures seem the more appropriate, the chief exceptions being issues of biogeography.

2.3.3 Rarity and range size estimates

As with abundances, this is not the place for a protracted consideration of the detailed methodology of mapping species' spatial distributions in order to obtain measures of range size. An extensive literature associated with this topic exists, though it is markedly more scattered than that relating to the estimation of population densities and population sizes (e.g. Udvardy, 1969; Taylor *et al.*, 1985; Haila *et al.*, 1989; Harrison, 1989; Smith, 1990; Harding, 1991; Bibby *et al.*, 1992a). With regard to the study of rarity, some general points should, nonetheless, be made:
(i)   The quality of mapping is a function of spatial resolution, abundance and region. The resolution at which species' ranges are mapped will be a

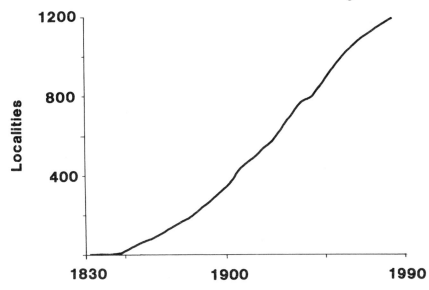

Figure 2.13 Cumulative number of new localities at which threatened taxa of vascular plants have been discovered in Finland from the mid-nineteenth century to the present. (Redrawn from Lahti *et al.*, 1991.)

compromise between that which enables geographic and environmental features to be resolved, and that at which patterns of sampling do not swamp patterns of distribution.

(ii)   As with abundances, great efforts may be necessary to map the ranges of rare species. Lahti *et al.* (1991), for example, show that the number of localities (geographically distinct populations) at which threatened plant taxa in Finland have been discovered has continued to rise from the early 1800s to the present (Figure 2.13). This is thought to result simply from increased research.

(iii)   The magnitude of the error associated with estimating the size of a species' range is unlikely to be constant for ranges of genuinely different sizes. Species which occur in fewer localities will tend to have their range sizes disproportionately underestimated, because any single locality contributes a greater proportion to a species area of occupancy and has a greater likelihood of contributing to its extent of occurrence. Russell and Lindberg (1988a) demonstrated this effect using data for marine molluscs. The known linear ranges along the coast were recorded for 180 species, and the overall distance spanned by all the species was divided into 10 km segments. By sampling these segments at random, 'observed' geographic ranges for different levels of sampling effort could be determined, from which a percentage error of estimation (PEE) of range size could be calculated (PEE = ((known range − 'observed' range)/ known range) × 100). As the number of sampling points increased, PEE decreased (Figure 2.14), and PEE remained significantly negatively correlated with known geographic range size even when as many as 50% of the segments were sampled.

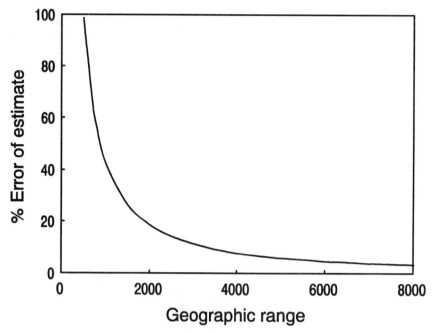

Figure 2.14 Relationship between the per cent error of estimate of geographic range size and known geographic range size (km) for 180 species of gastropods, when 50 (6%) of the possible 836 sample points are selected at random. (Redrawn from Russell and Lindberg, 1988a.)

(iv) The determination of a species' extent of occurrence is, in general, considerably easier than the determination of its area of occupancy. It is usually based on less information. However, the accuracy of a measure of extent of occurrence is often more prone to dependence on one or a few data points than is that of a measure of area of occupancy.

(v) When measuring species' extents of occurrence, it needs constantly to be borne in mind that range margins, in the sense of a strict border, do not as such exist (MacArthur, 1972). As Carter and Prince (1988) state with regard to plants, 'Distribution limits are simply lines drawn around species ranges on a map, and as such they are abstractions from reality. They do however relate to the ecology of real plants in dividing areas where they grow from areas where they do not.'

While the main body of occurrence is often differentiated from outliers, again such a distinction is usually arbitrary. Different workers generate maps in different ways and treat outliers differently.

(vi) Species which are regarded as rare by virtue of their small geographic range may in some instances have these ranges disproportionately well mapped. Occurrences of rare species at sites at which they have previously been unrecorded are apt to be noted. More notice may be taken of the occurrences of vagrant individuals of rare species compared with common ones, with a resultant disproportionate inflation in their range sizes. Likewise, because maps inevitably have a temporal dimension and are usually the summed records of at least several years, any tendency for rare species to turnover more frequently than common species (see Chapter 5) will lead to their recorded ranges being disproportionately large.

(vii) Species are seldom mapped with equal effort across their entire geographic ranges, some regions are better mapped than others. As a consequence, particular caution is required when comparing the levels and patterns of occupancy of different regions or the ranges of species which predominantly occur in different regions. For example, range maps of most taxa, even such well-known ones as birds, tend to show a fragmentation of the area occupied by individual species from east to west across Europe. While this pattern conforms with some ideas about the way in which the structure of ranges changes towards their limits (Chapter 4), any real effects are probably swamped by a strong east–west gradient in the quality of mapping. Significant strides are being made in reducing the problems of uneven and inadequate mapping of species occurrences, through the development of a variety of methods to smooth occurrence data and to generate predictions of the probabilities with which species are likely to occur in different places (Longmore, 1986; Bibby and Hill, 1987; Margules and Stein, 1989; Nicholls, 1989; Price and Endo, 1989; Hill, 1991; Osborne and Tigar, 1992b). Likewise, methods are becoming available for prediction of the abundances of individual species across large areas based on remotely sensed satellite imagery (e.g. Avery and Haines-Young, 1990; Rogers and Randolph, 1991).

2.3.4 One- , two- and three- dimensional ranges

Whether quantified as extents of occurrence or areas of occupancy, ranges are typically treated as at most two dimensional. This is, of course, a further simplification in trying to distil the spatial occurrence of a species into a single variable. In reality, the positions of individuals at any one time are defined with respect to three axes, not just one or two, and it is unlikely that positions on the z-axis (vertical, altitudinal or depth) will be equal for all individuals of a species. It can be argued that relative to the magnitude of the differences in the latitudinal and longitudinal positions of individuals, differences in altitudinal position will be minimal. While doubtless true of species distributed over tens of degrees of latitude or longitude, it is potentially a weak argument when dealing with rare species of limited distribution on these axes.

Although I have not encountered such an analysis, crude incorporation of the altitudinal dimension into quantification of a species' extent of occurrence would be comparatively straightforward. One could, for example, simply multiply one of the two-dimensional measures (such as the product of the latitudinal and longitudinal ranges, or the sum of the sizes of the geographic areas in which the species occurs; Table 2.1) by the altitudinal range over which a species occurs. Equally one might assess a species' area

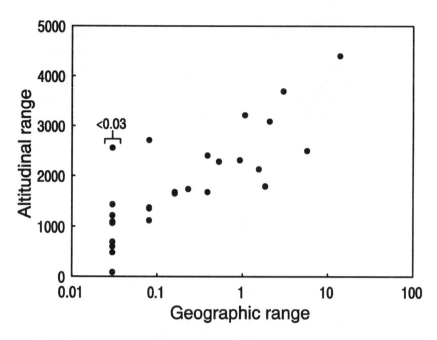

Figure 2.15 Relationship between the altitudinal range (m) and the geographic range ($\times 10^6$ km²) for North American species of *Peromyscus* (mice). (From data in Glazier, 1980.)

of occupancy on the basis of the numbers of cubes occupied in a three-dimensional grid.

It is at present difficult to assess what effect such considerations might have on our understanding of differences in species' range sizes. For a very few taxa we know that the latitudinal, or other horizontal planar, measure of range size and the altitudinal ranges of species tend to be positively correlated (Figure 2.15; e.g. Pielou, 1979; Glazier, 1980; Thomas and Mallorie, 1985a; Obeso, 1992; Stevens, 1992b), suggesting that the former measure may carry at least some information about the latter. Likewise, Graves (1988) found that the elevational range of Andean bird species and estimates of range width which accounted for changes in altitude were strongly positively correlated. Latitudinal and longitudinal ranges may also be correlated, as has been shown for North American terrestrial mammals, and North American and European land birds (Brown and Maurer, 1989). Further exploration of the interactions between ranges on the different dimensions and between one-, two- and three-dimensional measures of range size may prove very profitable. Until such time as these studies have been performed we will have to rely solely upon one- and two-dimensional measures of range size as simple descriptors of species' spatial distributions.

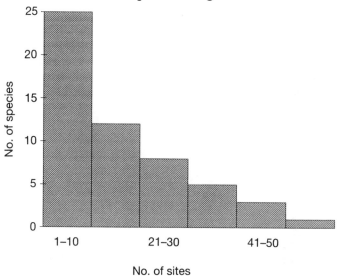

Figure 2.16 Number of sites (of a possible 62) at which species of ground beetles (Coleoptera: Carabidae) were caught in pitfall traps in the North York Moors National Park. (From data in Gardner, 1991.)

2.4   SPECIES' RANGE SIZE DISTRIBUTIONS

While species–abundance patterns have been subjected to many reviews and critical studies, the frequency distributions of species' range sizes have attracted considerably less attention. There have been no systematic attempts to compare the fits of different models nor to assess whether observed patterns differ between taxa or habitats.

At larger scales, macro-ranges and meso-ranges, the distributions are typically strongly right-skewed (Figures 2.16–2.19; e.g. Willis, 1922; Anderson, 1977, 1984a, b, 1985; Margules and Usher, 1981; Rapoport, 1982; Schoener, 1987; Russell and Lindberg, 1988a; Hengeveld, 1990; McLaughlin, 1992). A logarithmic transformation often normalizes these distributions, or at least makes them roughly symmetrical (e.g. Anderson, 1984a, b; McAllister *et al.*, 1986; Pagel *et al.*, 1991). Exceptions are not, however, infrequent (e.g. Anderson, 1977, 1984b), and a variety of other models has also been used. Buzas (Buzas *et al.*, 1982; Buzas and Culver, 1991) and others (Koch, 1987) have found, for example, good fits to a log-series model (Fisher *et al.*, 1943; Williams, 1964).

The differences observed may owe as much to the variety of ways in which ranges are measured, the variable quality of the data, and how different groups of species are treated (e.g. introductions, vagrants, species the ranges of which extend outside the region of study), as they do to genuine variation in patterns of spatial occurrence. In some data sets, skewedness is in part due to the inclusion of species, the entire ranges of which do not lie within the study area

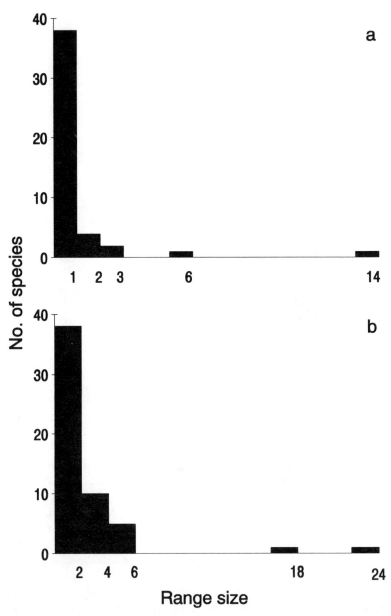

Figure 2.17 Frequency distributions of (a) the numbers of species of North American *Peromyscus* (mice) with geographic ranges of different sizes (extent of occurrence, $\times 10^6$ km^2). (From data in Glazier, 1980.) (b) The numbers of species of Nearctic *Nebria* ground beetles (Coleoptera: Carabidae) with geographic ranges of different sizes (maximum linear extent, $\times 10^3$ km). (From data in Kavanaugh, 1985.)

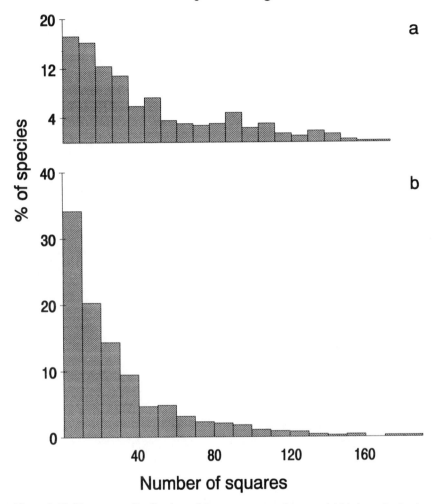

Figure 2.18 Frequency distribution of the percentage of terrestrial bird species having different sized geographic ranges in Africa, range being measured as the number of 2 ½° × 2 ½° squares in which the species were recorded. (a) Non-passerines, and (b) Passerines. (Redrawn from Pomeroy and Ssekabiira, 1990.)

(e.g. McAllister *et al.*, 1986). However, this bias is insufficient to explain departures from uniformity or from strict normality in those data sets where it does apply.

Debate as to why the distribution of species' ranges at large spatial scales takes the form it does has centred largely on Willis' (1922) age-and-area hypothesis. This states that the range of a species is indicative of its age, with

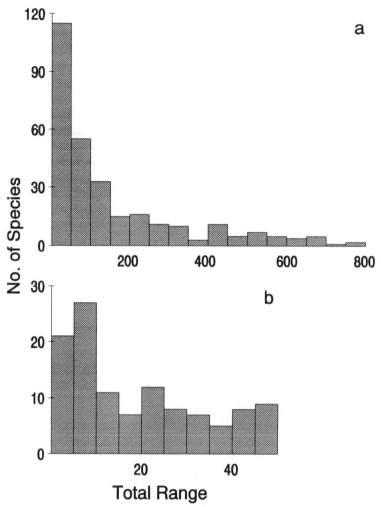

Figure 2.19 Frequency distribution of geographic range sizes (breeding and non-breeding) of Australian terrestrial passerines; (b) is a more detailed presentation of the first size class of (a). Ranges measured in quadrats of 10^4 km^2. (Redrawn from Schoener, 1987.)

younger species having smaller ranges. Discussion of the relationship between age and range size will be postponed until later (Chapter 5). The hypothesis has been challenged on several grounds (for references see Stebbins and Major, 1965; Fiedler, 1986). However, as McLaughlin (1992) observes, critics have not in general offered alternative explanations for the pattern.

Anderson (1985) used a Markov chain model to explore ways in which the 'hollow curve' might be generated, but without a better understanding of whether differences in observed patterns reflect biology or data collation, agreement or disagreement between a model and data is difficult to interpret.

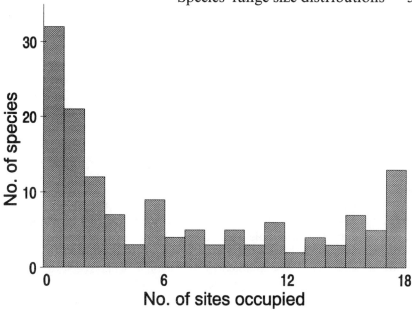

Figure 2.20 Frequency distribution of the number of watersheds (of a possible 18) occupied by plant species in the Konza Prairie Research Natural Area, Kansas. (Redrawn from Collins and Glenn, 1990.)

All in all, there is some temptation to suggest that geographic range sizes might best be described as logarithmically distributed and that they provide a further example of the central limit theorem, the distribution resulting from the interplay of many independent factors (see May, 1975).

Decreasing the area over which ranges are measured (i.e. moving from macro-ranges to meso-ranges and then to micro-ranges) tends to lead to the frequency distribution of their sizes becoming bimodal (Figure 2.20; Hanski, 1982a, b, c; Gotelli and Simberloff, 1987; Collins and Glenn, 1990, 1991). To the mode generated by species of limited range is added one generated by species which are more widespread. By convention, frequency distributions of species' meso-range and micro-range sizes are commonly constructed such that range size classes are defined in terms of the proportion of the possible range (usually a number of sites) at which species could occur. The right-hand mode may be generated by species which occur throughout the region of interest (the range size class, the upper limit of which is 100%), or by species which although the most widespread in the assemblage do not do so (a range size class, the upper limit of which is less than 100%). Which of the two forms of biomodality is observed seems to depend on the size of the areas over which ranges are observed. The smaller the areas, and the less heterogeneous the habitats they contain, the more likely the right-hand mode is to be produced by species which occur throughout the area.

The bulk of interest in biomodality has been in those cases in which the right-hand mode includes an upper limit of 100% occupancy. A number of explanations for such distributions of range sizes have been proffered. It has repeatedly been suggested that they may, in part at least, be sampling artefacts (Preston, 1948; Williams, 1950, 1964; McIntosh, 1962; Nee *et al.*, 1991a). Several such artefacts can contribute:

(i)   Species of low abundance have, regardless of the number of sites at which they occur, a low probability of being recorded at any one site because they are often more difficult to find. This will tend to inflate the numbers of species recorded as having small ranges and hence contribute to the left-hand mode of the frequency distribution.

(ii)   Vagrants usually occur at few sites, contribute most significantly to species' richness at meso-scales (see Chapter 1), and hence are likely to inflate the left-hand mode at this scale. Raunkiaer (1934) essentially made the same point, in arguing that biomodality was produced because species found to occupy all sites were those adapted to that habitat, while species found to occupy only a few sites were adapted to, and common in, other habitats.

(iii) The largest range size class of the frequency distribution may be larger than for other classes, leading to the right-hand mode of the distribution. This may simply be because all species whose ranges are greater than the lower limit to the class are placed in it regardless of how much greater they are. Thus, the data do not necessarily show biomodality at all.

(iv)   If the individuals of each species are dispersed randomly in space and independently of other species, then a bimodal distribution of occupancy will be generated when the inter-specific abundance distribution accords to a log-series or a log-normal (Preston, 1948; Williams, 1950, 1964). This is because the higher classes of occupancy include a wider range of abundances. Individuals are seldom likely to be randomly distributed in space except at the very smallest scales (Chapter 3). Nonetheless, Williams' (1950, 1964) model provides a reasonable fit to some regional data (Hanski, 1982c).

In the belief that it is not simply a sampling artefact, various other explanations for bimodality have been suggested. In the main these have stemmed from work on the dynamics of metapopulations, and in particular from developments of Levins' (1969) metapopulation model. This model expresses the rate of change in the proportion of sites which are occupied as an outcome of the difference between the rate at which previously empty sites are colonized (i.e. the immigration rate) and the rate at which previously occupied sites become unoccupied (i.e. the extinction rate). While in its original form it generates unimodal frequency distributions of occurrence (most species occur at few sites and a few occur in most sites), some modifications result in bimodality (Hanski, 1982a, 1991a; Gotelli, 1991). Of these, the core–satellite hypothesis (Hanski, 1982a, b, c) has received considerable attention (species forming the right-hand mode are termed 'core', while species forming the left-hand mode are termed 'satellite'). Whereas Levins' original model assumes that the rate of colonization of patches is dependent upon the proportion of patches already occupied and the rate of extinction is not, the core–satellite model assumes that both rates depend on the proportion of patches already occupied. The core–satellite model has been subjected to a number of empirical tests and to some conceptual challenges (Hanski, 1982a, b, c; Gotelli and Simberloff, 1987; Williams, 1988; Gaston and Lawton, 1989; Collins and Glenn, 1990, 1991; Nee et al., 1991a; Obeso, 1992). It seems to have weathered these quite well. However, it does have problems. In particular, it lacks realism in that rare species cannot go extinct from the entire patch system, and it predicts that species should through time switch between core and satellite status. These difficulties seem set to be overcome by new generations of models (Hanski, 1991a).

An unfortunate complication to interpreting the validity of the core–satellite hypothesis has been the widespread application of the terms core and satellite, with little regard to whether data concord with the assumptions and predictions of the model. Indeed, there are examples of bimodality in the frequency of species' occurrences for which explanations based on metapopulations are likely to be entirely inappropriate. Metapopulation models strictly only apply to regional patterns, and not within-community patterns, where sampling effects are more likely to provide appropriate explanations for bimodality. Many of the models do not apply to situations in which distances are too great for there to be movement of individuals from one site to the majority of the rest. Explanations for bimodality in these instances remain unsatisfactory.

2.5 PSEUDO-RARITY AND NON-APPARENT RARITY

Solely as a consequence of the differing reliability of estimates of their abundance or range size, some species in an assemblage may be regarded as rare when they are not, and others may not be regarded as rare when in fact they are. These two states can be referred to as pseudo-rarity and non-apparent rarity, respectively. Under some definitions of rarity one state cannot occur independently of the other. For example, under a proportion of species definition (Chapter 1) the misclassification of a species as rare when it is not must necessarily entail the misclassification of a species as not being rare when in fact it is.

In addition to artefactual impressions of species' relative abundances or range sizes, generated as a direct result of the real magnitude of these traits, there are a host of other mechanisms by which pseudo-rarity and non-apparent rarity can be generated. In the main these do not trouble studies of large-bodied temperate taxa.

2.5.1 Taxonomic complications

The failure to discriminate the specific identities of individuals reliably may result in the inflation of the abundances or range sizes of some species and the underestimation of those of others. As with most causes of pseudo-rarity and non-apparent rarity this problem is particularly acute in work on tropical faunas and floras. Thus, Gentry (1992) writes, 'Unless a tropical genus or family has been recently monographed, many of its apparently endemic species often turn out to represent taxonomic artifacts resulting from careless or parochial taxonomy rather than true endemics.'

2.5.2 'Inappropriate' sampling techniques

Species may appear to be rare because methods appropriate to determining their true abundances and range sizes have not been applied, have not been applied in the right place, or have not been applied at the right time. Schmidt and Buchmann (1986) provide a salutory example. Mutillid wasps, while brightly coloured, are seldom observed in large numbers and it is generally assumed that their population densities are very low. In the course of studies of a species of bee, these authors placed emergence traps over its nesting aggregations. These yielded large numbers of the parasitic mutillid wasp *Dasymutilla foxi*, suggesting that it occurred at a density of 76 individuals/m². In 6 years of investigating the bee species the authors had never seen *D. foxi*. They suggest that the species is active at times other than when they were present, and is behaviourally cryptic.

In a similar vein, adults of the ichneumonid *Diacritus aciculatus* will be regarded as exceedingly rare, even in optimal habitat, unless sampling is carried out during the narrow window of 10 days or so in early summer when they can be caught in large numbers (I.D. Gauld, personal communication). Espadaler and López-Soria (1991) suggest that rarity in Mediterranean ant species may be explained by insufficient sampling in time of adequate microhabitats or by the inconspicuousness of social parasitic species.

Disney (1987) suggests that, in general, too limited a selection of collection methods are employed in mapping the occurrence of insect species, with a resultant underestimation of their distributional extent. An obvious problem occurs when methods appropriate to the sampling of a species are unknown.

2.5.3 False assumptions

The abundance and distribution of species may be underestimated simply because incorrect assumptions have been made about these variables. Harding (1991) provides an example, 'the scarce emerald damselfly *Lestes dryas* was regarded as extinct in Britain (but not in Ireland) between 1973 and 1982, but was 'refound' in 1983 and has since been recorded in over 18 10-km squares in Britain. In all probability it survived in several areas between 1973 and 1982, but was not sought at new sites because it had genuinely died out at some previously well-known sites.'

Conversely, it is possible for very abundant species to be overlooked; Gentry (1992) gives examples of several common species of tropical plants which have only very recently been described.

What we perceive as pseudo-rarity and non-apparent rarity constitute a hindrance to studies of rarity, and more generally of population and community structure. They may, nonetheless, be adaptive for the organisms concerned. An ability to avoid discovery is a good means of avoiding being consumed. Equally, an ability to appear more abundant than is actually the case may be tantamount to ensuring that conspecifics encounter one another. Indeed, for individuals of any given species the tension between avoiding discovery and wanting to be found may play a profound role in their ecology.

2.6 PHYLOGENETIC RELATIONSHIPS

Rarity is a species' characteristic, although not a species-specific one because it is context dependent (Chapter 1). Like other species' traits it is thus likely to be a product of phylogenetic constraint as well as of independent adaptation. Closely related species tend to have more similar biologies than more distantly related ones, and thus are also likely to have more similar abundances and range sizes. The relative importance of phylogenetic constraint and independent adaptation in determining a species' abundance and range size is a significant methodological and heuristic consideration in the study of rarity.

This said, by and large, studies of rarity have taken little account of phylogenetic relatedness. Those which have, serve to highlight aspects of its potential importance. Maurer (1991) reports that 41% of the variance in the population densities of North American birds could be attributed to differences in the family to which they belonged. Hubbell and Foster (1986) test the numbers of species in different abundance categories in individual genera, against the distribution of all species in their study among those categories, and find genera for which the probability that all species would by chance be so rare or so common is extremely unlikely.

Failure to account for phylogenetic contributions can seriously bias statistical tests in inter-specific comparisons (Harvey and Pagel, 1991). In particular, it leads to a tendency towards the overestimation of the numbers of available degrees of freedom, because species are incorrectly treated as independent data points. Thus the significance levels for statistical tests tend to be inflated. The consequences of ignoring phylogenetic relatedness can be profound. In an extreme example, Gregory *et al.* (1991) found that using cross-species regression (without controlling for phylogenetic effects) among 75 tests performed they could detect 31 significant correlates between host ecology and host life history and the richness of cestode, nematode and trematode parasites of birds. Applying the technique of evolutionary covariance regression to control for host phylogeny, the number of significant correlates fell to three.

Unfortunately, the paucity of studies of rarity which consider the effects of phylogeny is difficult to rectify quickly. Reliable estimates of phylogenies, as opposed to taxonomies, are available for few of the taxa for which studies of the causes and correlates of rarity have been performed. Since both a reanalysis of published work or major new analyses lie beyond the scope of this volume we are reliant upon the existing literature. It seems reasonable to assume that were the phylogenetic effects to be removed many documented relationships would be found to be substantially weaker than they seem at present. At the risk of some later regret, I suggest, however, that much of the discussion in this book will be little altered by accounting for phylogenetic effects.

2.7 CONCLUDING REMARKS

The potential obstacles and pitfalls to the measurement of rarity have been emphasized in this chapter, somewhat at the expense of examples of what can be achieved. In no way should this be read as an indictment of quantitative approaches to the topic. Rather, it is meant to stimulate an awareness of the problems and some of the steps which can be taken to avoid them, in the hope that more rather than less study might result. As will be revealed in subsequent chapters, the outcomes of many studies pertaining to rarity are difficult to interpret because potentially they fall foul of some of the difficulties highlighted. This is not, however, their primary fault. Rather, it is the failure to provide adequate information by which the likely severity of these problems can later be assessed.

The remainder of this book is primarily concerned with developing an understanding of rarity on the basis of those studies which have been performed thus far, trying to take account of their strengths as well as their weaknesses.

3 The non-independence of abundance and range size

Thinking means connecting things, and stops if they cannot be connected.

G.K. Chesterton (1909)

In considering the characterization and measurement of rarity (Chapters 1 and 2) it has proved convenient to treat as separate the abundance of a species and its range size. This is, nonetheless, misleading. Abundance and range size are not independent. At its crudest this is merely to state that at least one individual of a species must be present for a quadrat, site, or region to be correctly recorded as part of its area of occupancy, or as extending the limits of its extent of occurrence. The range of a species is a result of spatial variation in its abundance, with the size of the range reflecting the presence/absence component of this variation. The objective of this chapter is to explore the relationship between abundance and range size. The chapter divides broadly into two parts, concerned, respectively, with intra-specific and inter-specific interactions. The former, although in the main not explicitly about rarity, provides some important background for the latter.

3.1 LEVELS OF RESOLUTION

As discussed in the previous chapter, the size of a species' range in terms of its area of occupancy will depend on the resolution at which the spatial occurrence of its individuals is mapped. In general, the more refined the scale of mapping, the smaller will be the observed range size. This is because it becomes possible to discriminate more and more areas in which the species does not occur and hence to exclude these from the measured area. Ultimately, of course, one can envisage having a map of the positions in space of each individual of the species in the region of interest, perhaps at a given instant in time or averaged over a defined period. The range size might then, in one sense, equate to the abundance of the species.

The decline in area of occupancy with increasing resolution is reminiscent of the problem of measuring the length of a piece of coastline. Such a measurement is made by counting the number of times, N, a 'ruler' of length l can be stepped around the boundary on a map. The length of the coastline is the product of N and l. As l is reduced, the steps reflect more of the detailed structure of the coastline and the resultant measure of its length increases. In fact, in the limit this length can be regarded as infinite. However, as Mandelbrot (1967, 1982) demonstrated, a constant, D, can be found, such that:

$$E = Nl^D \qquad (3.1)$$

where E is the fractal extent, and D the fractal dimension of the coastline. A smooth differentiable curve would have a fractal dimension of 1; increasingly complex boundaries would have dimensions which tend toward 2.

In principle, the area of a species' range can be treated in a similar fashion to the length of a coastline. The rate of decline in the area occupied with increasing resolution indicates the fractal dimension.

Whether species' ranges can be described in terms of fractal geometry (across all or a given span of spatial scales) remains to be determined. As yet there are several obstacles to doing so. Most significantly, analyses are hampered by the fact that with increasing resolution real declines in measured areas of occupancy are paralleled by declines in sampling effort. Species' patterns of occurrence can readily be mapped at a variety of resolutions, for example, published maps exist of the spatial occurrence of species of various taxa in Britain on grids from 1×1 km to 10×10 km (e.g. Heath *et al.*, 1984; Barnham and Foggitt, 1987; Plant, 1987), and the latter can readily be blocked up to yet coarser resolutions. Thus, if means can be found of overcoming the sampling inequalities, some of the basic data on occurrence may already exist in an appropriate form.

Notwithstanding these problems, an interesting feature of maps of the occurrence of a given species at different levels of resolution is that they often tend to show similar kinds of patchiness. Perhaps the classic example is provided by Erickson's (1945) maps of the distribution of *Clematis fremontii* var. *riehlii* in the Ozarks, at the level of counties, regions within counties, local areas (clusters) within regions, local habitats (glades) within clusters, and of all individuals within an aggregation within a glade (Figure 3.1). Pairs of these maps appear, subjectively at least, to bear some of the hallmarks of a fractal pattern (Williamson and Lawton, 1991).

If one takes to its logical conclusion the idea that a species' range is fractal over all levels of resolution down to the individual, then there is in fact no definable area to the range. At the limit, the area would at any one moment be the sum of the ground area occupied by each individual organism.

The occurrence of the individuals of a species needs a set of three coordinates to specify their position in space (longitude, latitude and altitude, or their equivalents). This raises the additional possibility of analysing their occurrence patterns in a similar way to that in which the spatial distribution of galaxies has been studied (Frontier, 1987). Here, one would be interested to determine the rate of change in the volume occupied with increasing resolution.

In the present context, the importance of these considerations perhaps lies not so much in whether the ranges of species do ultimately prove to be fractal or not, but that they emphasize that the abundance and the area of occupancy, at a given resolution, of a species both lie on the same continuum. The

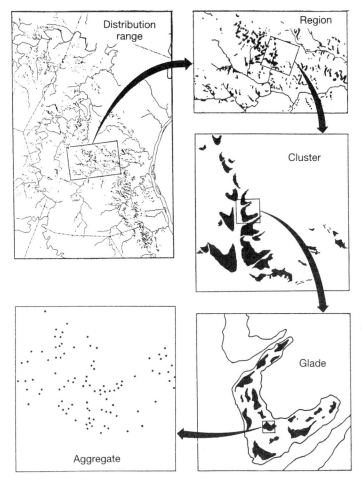

Figure 3.1 The patterns of occurrence of *Clematis fremontii* var. *riehlii*, at a hierarchy of scales. (Reproduced from Erickson (1945) with the kind permission of Missouri Botanical Garden.)

consequences of fractal form would perhaps, however, mean that the pattern of a species' spatial occurrence measured at one level of resolution could readily be used to predict that at other levels. As well as of innate heuristic interest, this would have important practical implications. It will only ever be possible to map the spatial occurrence of the majority of species in a very coarse fashion and an ability to estimate their occurrence at finer resolutions might provide a valuable management tool. This is especially true of rare species, for which mapping is frequently very demanding of resources (Chapter 2). In particular, fractal descriptions of species patterns of occurrence might usefully be combined with probabilistic approaches to reducing the limitations of inadequate mapping (e.g. Osborne and Tigar, 1992b).

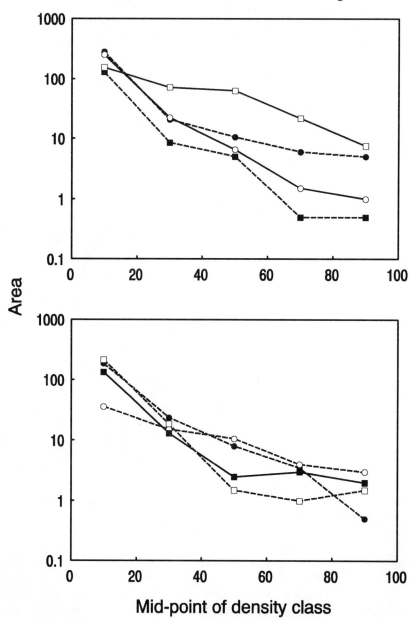

Figure 3.2 The area of their geographic ranges in North America over which eight species of wrens (*Troglodytidae*) reach different densities. Density classes are 0.5–20%, 20–40%, 40–60%, 60–80% and 80–100% of the maximum local abundance (maximum local abundance is the average value that was greater than 99% of all the values for a given species for the number of individuals seen per party-hour of counting). (From data in Root, 1988.)

3.2   DENSITIES AND PRESENCE/ABSENCE DATA

The individuals of a species are not uniformly distributed in space. They occur in greater numbers in some places than in others. If levels of occupancy are expressions of levels of abundance, it thus follows that differences in abundances will translate into differences in levels of occupancy. Indeed, for a variety of types of statistical distribution of entities among sampling units, the proportion of sampling units in which those entities are recorded and their mean abundance across those units are formally related. Even were individuals to be distributed at random in space, there would be a positive correlation between the proportion of sampling units in which they were recorded and their mean abundance across those units; incidence and density are formally related with respect to the Poisson probability distribution.

The Poisson distribution provides an adequate description of the spatial occurrence of individual animals or plants among sampling units only at very small sampling scales or perhaps when the species concerned only exists at exceedingly low abundances. Over most scales individuals tend to be aggregated in their spatial occurrence (Taylor *et al.*, 1978). Indeed, as with inter-specific frequency distributions of abundance (Chapter 2), particular species tend to occur in most places at comparatively low densities and in a few at relatively high densities. The frequency distribution of within-species' abundances is long-tailed (Figures 3.2–3.4); the same also tends to be true of population sizes (Barrett and Kohn, 1991). The negative binomial is widely regarded as the best descriptor of the pattern of spatial occurrence of individuals of a species at a given point in time, although various alternatives have been proffered (e.g. Perry and Taylor, 1985).

For populations in the field, the negative binomial parameter k (often referred to as the clumping parameter) is related, non-linearly, to population density (Perry and Taylor, 1986). However, a general empirical relationship between the proportion of samples in which a species has been recorded and its mean density has been established, such that:

$$\log[-\log(\Omega_o)] = \log\alpha + \beta\log\mu \tag{3.2}$$

in which Ω_o is the proportion of unoccupied samples, μ is mean density, and α and β are constants (Nachman, 1981, 1984; Kuno, 1986; Perry, 1987; Yamamura, 1990)

Interest in the intra-specific relationship between frequency of occurrence and density has largely been stimulated by economic considerations. Counting the number of sampling units in which a species is found and using this information to calculate its density is more cost-effective than determining that density directly. It thus provides an efficient means of monitoring the populations of pest species for the purposes of, for example, choosing the optimal time to spray pesticides. One consequence of this motivation has been that most considerations of intra-specific relationships between density and

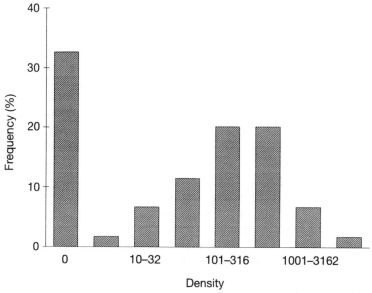

Figure 3.3 Frequency distribution of the mean adult population density (adults/m^2) of the zebra mussel (*Dreissena polymorpha*) in each of 57 lakes within the European range of the species. (Redrawn from Strayer, 1991.)

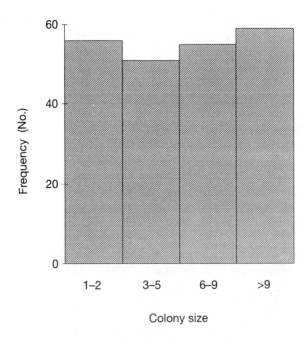

Figure 3.4 The number of heronries of different sizes (number of nests) for the grey heron (*Ardea cinerea*) in Scotland, as counted in the 1985 British Trust for Ornithology census. (From data in Marquiss, 1989.)

occurrence have concerned sampling units of fine resolution, such as individual leaves or plants (e.g. Gerrard and Chiang, 1970; Wilson and Room, 1983; Ward *et al.*, 1986; Hergstrom and Niall, 1990; Yamamura, 1990).

Information on intra-specific relationships between levels of abundance and occurrence has been less forthcoming for coarser spatial resolutions. Their possible applications here have, nonetheless, not passed unnoticed. In particular, the possibility of using information on the numbers of sites at which particular bird species have been recorded to monitor their abundance is attractive. This is because, as alluded to earlier, the amateurs who carry out most of such survey work tend to prefer to compile lists of species than to generate counts of their abundances.

Bart and Klosiewski (1989) evaluated the similarity between changes in the populations of North American birds estimated by changes in their densities and by changes in their presence or absence from sites. They found that the two paralleled one another quite closely (Figure 3.5), with the proviso that, unsurprisingly, changes in the numbers of stations occupied were smaller than changes in the numbers of individuals. The suggestion has been made that changes in species' occurrence may be especially valuable in monitoring the population changes of species which have very low abundances (rare species) and for which trends cannot be estimated more directly (Robbins *et al.*, 1989).

One consequence of a strong relationship between a species' abundance in an area and its range size there is that, as with abundances, the frequency distribution of the micro- or meso-range sizes of a species at different points in space tends to be strongly right-skewed, such that in most places it is restricted in occurrence, while in a few it is widespread (Figure 3.6).

3.3   INTER-SPECIFIC ABUNDANCE AND RANGE SIZE

The existence of intra-specific abundance–range size relationships plainly begs the question as to whether there is an equivalent inter-specific relationship. That is, do species which are rare in terms of their abundance have a high probability of also being rare in terms of the size of their range.

There is no reason why intra-specific relationships between species' abundances and range sizes should translate into inter-specific relationships between these variables. Nonetheless, intra- and inter-specific patterns of rarity commonly parallel one another, and confidence in the existence of an inter-specific relationship has been sufficient for a number of studies to employ data on species' presence/absence at moderate resolutions as measures of their abundance relative to other species (Neuvonen and Niemelä, 1981; Fowler and Lawton, 1982; Niemelä and Neuvonen, 1983; Kennedy and Southwood, 1984). Fowler and Lawton (1982) and Kennedy and Southwood (1984), for example, use the number of tetrads in which plant species have been recorded as measures of their local abundances, and the number of 10 km squares in which they have been recorded as measures of their range sizes. In both

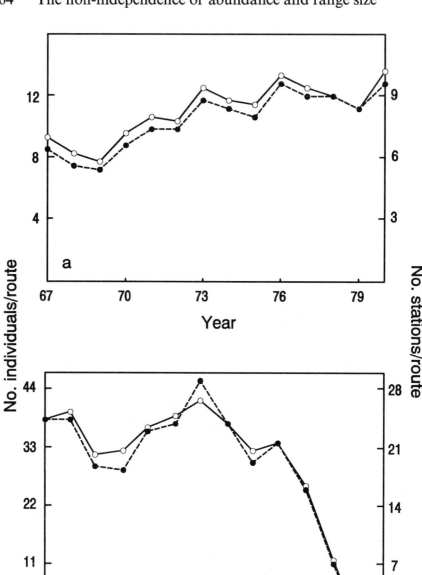

Figure 3.5 Changes in the abundance of the northern cardinal *Cardinalis cardinalis* estimated using numbers of individuals (solid circles) and numbers of stations at which the species was recorded (open circles). (a) Across years, and (b) along a longitudinal gradient. (Redrawn from Bart and Klosiewski, 1989.)

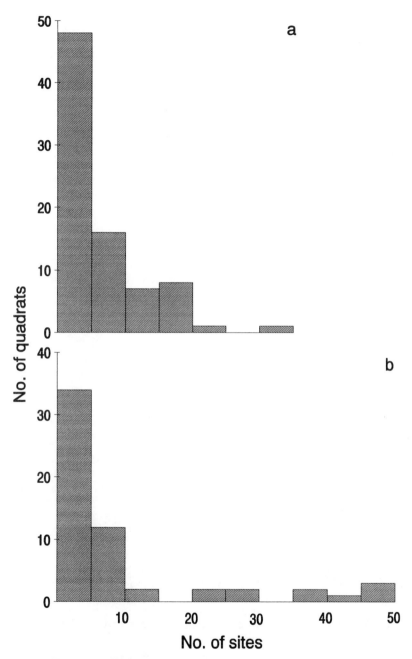

Figure 3.6 Frequency distributions of the numbers of quadrats containing different numbers of sites at which a species was recorded, for (a) the common toad (*Bufo bufo*) in Sweden (quadrats 1° × 1°) and (b) heather (*Calluna vulgaris*) in Dorset, Britain (quadrats approximately 4 × 4 miles). (From data in Gislen and Kauri, 1959; and Good, 1948.)

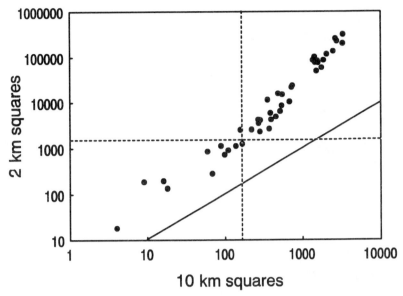

Figure 3.7 Relationship between the range size in 2 km (tetrads) and 10 km squares of umbellifer species in Britain. Solid line defines the lower bound to the possible relationship. (From data in Fowler and Lawton, 1982.)

instances the two measures are strongly positively correlated across species (Figure 3.7).

3.3.1 Patterns

In general, faith in the existence of a broad inter-specific abundance–range size relationship in situations where it cannot be demonstrated, seems fully justified. Positive relationships between the within-range abundance and the range size of species within an assemblage have been widely documented (Figures 3.8–3.10; Williams, 1964; Hanski, 1982a; Brown, 1984; Söderström, 1989; McCoy, 1990; Kouki and Häyrinen, 1991; Obeso, 1992; see Gaston and Lawton, 1990a for additional references). More widespread species tend to have greater abundances. Put another way, species whose abundances lead to them being categorized as rare are also likely to have range sizes which would lead them to be classified as rare. This broad relationship tends to hold despite a number of variations in the way analyses are performed.

- Abundances may be means or maxima (e.g. Bock and Ricklefs, 1983).
- Abundances may be from one site in the range (e.g. Owen and Gilbert, 1989; Gaston and Lawton, 1990a), averaged across several sites from one part of the range or averaged across several sites spread throughout the range (e.g. Bock and Ricklefs, 1983; Bock, 1984; Brown and Maurer, 1987). The last of these is the more usual form of analysis, and equates most closely to the intra-specific relationships discussed earlier.

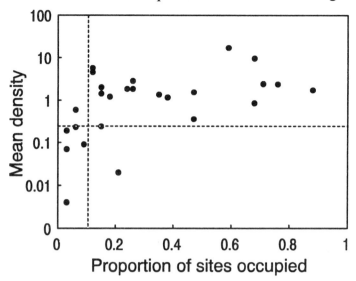

Figure 3.8 Relationship between the average population density (per cm²) when present, and the frequency of occurrence (number of lakes occupied/34) of species of crustacean zooplankton in 34 lakes in north western Ontario. (From data in Wright, 1983.)

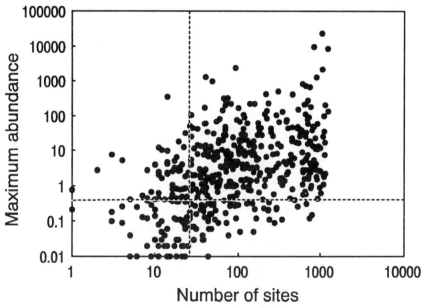

Figure 3.9 Relationship between the maximum local abundance value of North American birds and the number of sites at which each was recorded (maximum local abundance is the average value that was greater than 99% of all the values for a given species for the number of individuals seen per party-hour of counting). Dashed lines delineate those species which under a quartile definition are categorized as rare. (From data in Root, 1988.)

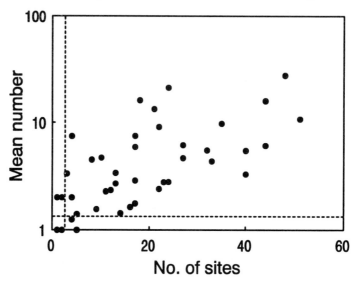

Figure 3.10 Relationship between the mean number of individuals when present, and the number of sites at which recorded, for species of ground beetles (Coleoptera: Carabidae) captured in pitfall traps at 62 sites in the North York Moors National Park. Note, several data points overlie one another. Dashed lines delineate those species which under a quartile definition are categorized as rare. (From data in Gardner, 1991.)

- Range sizes may be measured as extents of occurrence (e.g. Brown and Maurer, 1987) or areas of occupancy (e.g. Hanski, 1982a; Gaston and Lawton, 1988a, b; Ford, 1990).
- Range sizes may be measured at meso- (e.g. Hanski, 1982a; Collins and Glenn, 1990; Gaston and Lawton, 1990a) or macro-scales (e.g. Brown and Maurer, 1987; Ford, 1990). Although they will not be considered further here, there are also examples of positive abundance–range size relationships for micro-scales (e.g. Collins and Glenn, 1990; Wright, 1991). Bock (1987) demonstrated, for Arizona land birds, that positive abundance–range size relationships can be recovered for the same assemblage at different spatial scales (Figure 3.11). Rapoport *et al.* (1986) reported that the abundances of plants locally were positively correlated with their range sizes at two scales.
- The taxa considered may differ greatly (plants – Gotelli and Simberloff, 1987; Collins and Glenn, 1990; plankton – Wright, 1983; insects – Hanski, 1982a; Gaston, 1988; Williams, 1988; Kemp *et al.*, 1990; McCoy, 1990; fish – Macpherson, 1989; birds – Bock and Ricklefs, 1983; Bock, 1984, 1987; Brown and Maurer, 1987; Ford, 1990; Gaston and Lawton, 1990a; mammals – Brown, 1984). The observed relationship does, however, tend to weaken as the relatedness and ecological similarity of species in an assemblage declines. Indeed, the effects of the taxonomic relatedness of the species

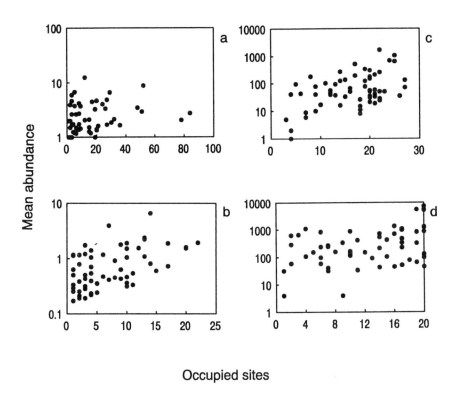

Figure 3.11 Relationships, at four spatial scales, between the mean abundance within occupied sites and the number of sites occupied for 62 landbird species wintering in southeastern Arizona. (a) 124 35 m radius plots in the Huachuca Mountains (abundance is the average number counted per occupied plot cumulatively during five censuses of each plot). (b) 24 habitat types in the Huachuca Mountains (abundance is the number counted cumulatively during five censuses of each plot, averaged over all plots in each occupied habitat type). (c) 27 Arizona Christmas count circles (24 km diameter; abundance is the average number counted per 100 party-hours of field effort in each occupied circle). (d) 20 5° blocks across the USA west of the 100th meridian (abundance is the average number counted per 1000 party-hours of field effort in each occupied block).(From data in Bock, 1987).

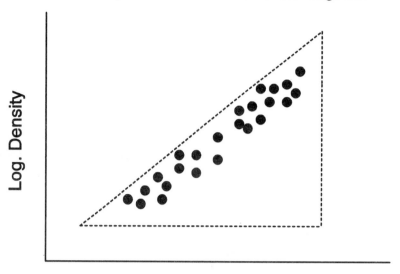

Log. Range size

Figure 3.12 Generalized figure of the inter-specific relationship between population density and range size. With increasing numbers of species, decreasing relatedness and increasing spatial scales, points tend to fall within the dashed triangle, rather than forming a simple linear trend as shown.

comprising an assemblage on the observed interaction between abundance and range size could usefully be investigated in some depth.

The amount of variation explained in such analyses is highly variable. Particularly when assemblages are speciose and range sizes are measured over macro-scales it may be very low. In such instances, data points tend to fall within a roughly triangular area of the abundance–range size plot, with maximal density tending to increase with range size, there being no species with small ranges and high densities, and there being species of both high and low densities with large range sizes (Figure 3.12). Here, it may be most interesting to ask what determines the broad constraints on realized combinations of abundance and range size, rather than what determines any weak correlation between the two (Brown and Maurer, 1987).

There are a few examples of negative relationships between abundance and range size, and of instances in which no significant relationship could be detected (negative – Arita *et al.*, 1990; Gaston and Lawton, 1990a; no relationship – Thomas and Mallorie, 1985b; Spitzer and Lepš, 1988; Arita *et al.*, 1990; Gaston and Lawton, 1990a, b). Thus, for example, Arita *et al.* (1990) found a weak negative correlation between density and geographic range size for 100 species of Neotropical mammals. While the relative infrequency of relationships which are not positive may be grounds for ignoring them as chance events, the ease with which hypotheses can explain such cases may prove a useful test of their realism (Gaston and Lawton, 1990a; Hanski *et al.*, 1993).

3.3.2 Possible explanations

Six types of possible mechanism can be proffered as explanations for, or contributors to, the positive relationship between abundance and range size. As yet there has been little attempt to compare the explanatory power of each for any one data set. Ultimately this may prove a necessary step to distinguish between them.

(a) Patterns of aggregation

For meso-scales, one can visualize the data point for each species in an inter-specific abundance–range size plot as lying on a separate intra-specific relationship. In other regions, a species would have a higher level of occupancy where it was more abundant and a lower level of occupancy where it was less abundant. A variety of constraints upon the slopes of these intra-specific plots could result in an inter-specific abundance–range size relationship being observed. That is, the inter-specific relationship could be explained in terms of some general tendencies for the levels of occupancy of individual species to increase with their mean abundance.

Such a hypothesis is difficult to test without a great deal of information on the real patterns of aggregation of individual species with respect to changes in mean density.

In one sense an explanation based on patterns of aggregation serves solely to move the level of explanation. The question of why the regional or global population of a species is distributed in a particular way remains.

(b) Artefacts

There are several ways in which an artefactual basis to abundance–range size relationships might arise.

● For any two species having equal sized ranges, that with the lower abundance is likely to be recorded as occurring at fewer sites or in fewer regions, simply because it has a lower probability of being found there. This sampling artefact will generate a spurious, and potentially strong, positive correlation between abundance and range size. The existence of such an effect has been widely noted (Brown, 1984; Gaston and Lawton, 1990a; Wright, 1991; Hanski *et al.*, 1993), but means of effectively partitioning it out of abundance–range size analyses have largely remained elusive. The difficulty lies in the need to make assumptions about the real spatial distribution of the abundances of each species in order to do so. Simple models can be applied, but without knowing how appropriate they are, conclusions are difficult to interpret. It is important to recognize the distinction between an explanation of the inter-specific abundance–range size relationship based on the real spatial patterns of aggregation of the individual species (section (a) above), and one based on the effects of the inadequacies of sampling. Intra-specific relationships between abundance and range size underpin both. In the former case, however, the argument concerns relationships between the abundance and occupancy of individual species, in which changes in abundance and occupancy result from real changes in abundance and occupancy either in time or space. In the second case it concerns relationships in which changes in observed occupancy are a product of increasing the numbers of individuals which are sampled; the abundance and occupancy of the species concerned has not itself changed.

● Presence/absence information is sometimes substituted for estimates of density when densities are very low. This will inevitably tend to generate an artefactual component to abundance–range size relationships.

● If, when abundances are averaged across a series of sites or areas, zero values are included, then clearly species which have smaller ranges and hence occupy fewer sites will tend to appear to be less abundant. This will tend to generate a positive bias to abundance–range size relationships. Lacy and Bock (1986) demonstrated that accounting for such an effect weakened the observed correlations between abundance and range size.

● If range sizes are measured in terms of the frequency of occupancy of small insular areas or sites (e.g. true islands), rare species may be absent from some of these for purely probabilistic reasons. Haila and Järvinen (1983) find that 75% of the breeding land bird species which were absent from the island of Ulversö were rare on the larger island of Main Åland, and that many of these absences can be explained by differences in land area (Table 3.1). Areas of Main Åland of the same size as Ulversö would not be expected to support a greater number of rarities than the latter does.

Table 3.1 The numbers of breeding land bird species absent from Ulversö but found on Main Åland, and the reasons for their absence. (From Haila and Järvinen, 1983)

	Short term	Long term
Rarity	40	17[a]
History/dispersal	1.5	4.5
Habitats unsuitable	6.5	25.5
Inter-specific competition	1	1
Predation (of nests)	2	3
Obscure reasons	2	2

[a] In this column rarity is not a cause of absence; these species are expected to breed on Ulversö, albeit seldom.

● Differences in the detectability, or conspicuousness, of species could lead
  to positive abundance–range size relationships independent of differences
  in their real abundances. Species which were easily observed would tend to
  have greater recorded abundances and range sizes than those which were
  more difficult to find. The sole test of such an effect, for woodpeckers and
  passerines, found none (Bock, 1987). Further studies are needed to ensure
  the generality of this conclusion. It is very likely that even if insufficient to
  explain a significant amount of the between-species variation in abundances and range sizes, detectability does play a role in determining the
  relative magnitudes of recorded abundances and range sizes. One can
  imagine, for example, that the abundances and range sizes of some soaring
  birds, which can be readily seen over large distances, might be better
  recorded or even overestimated in comparison with those of secretive
  skulking species. Indeed, it is not unknown for the abundances of large and
  very obvious species to be seriously overestimated.

(c) Range position

If it cannot be explained as simply a sampling artefact or a result of patterns
of aggregation, the abundance–range size relationship may be a product of
the abundance structure of the geographic ranges of the species (see Chapter
4). This might be the case if species' abundances and areas of occupancy
decline toward their range limits. Species at their limits in the region of study
would have low abundances and small ranges, while those toward the centre
would have higher abundances and larger ranges. This effect plainly cannot
explain abundance–range size relationships based on the entire geographic
ranges of species, and has consciously been controlled in some analyses (e.g.
Bock, 1984). Nonetheless, it may well contribute to many documented patterns, depending upon the proportion of between-species variation in abundances which can be accounted for simply by the position of the area in which
those abundances were measured with respect to the different geographic
ranges of the species. In some systems, a high proportion of the species
classified as rare are at the limits of their geographic ranges (Table 3.2).

(d) Vagrancy

The fourth explanation of abundance–range size relationships relates to the
possible effects of vagrant individuals. It could simply be that species with
small ranges are disproportionately more likely to occur as vagrants, because
the probability of individuals entering an area which is not part of a species'
true geographic range is that much greater than for widespread species. This
would increase the number of areas in which a narrowly distributed species is
recorded at low abundances and thus decrease its mean abundance.

As far as I am aware there is no evidence for such a mechanism, though this
would be unsurprising given the difficulty of differentiating vagrants from

Table 3.2 The percentage of plant species of differing levels of occurrence and habitat affinity in the Sheffield region which are at the limits of their geographical range in Britain. Figures in parentheses are the percentage within each grouping at the northern edge of their range in the Sheffield region. (From Hodgson, 1986a.)

	Extant flora						Extinct flora	
	Total		Common		Rare			
All habitats	16	(61)	2	(33)	58	(66)	56	(88)
Aquatic	16	(80)	0		27	(100)	43	(100)
Mire	11	(68)	0		33	(70)	42	(75)
Arable	13	(100)	0		80	(100)	77	(100)
Open	19	(50)	2	(0)	67	(56)	89	(75)
Grassland	15	(50)	4	(50)	58	(57)	37	(100)
Woodland	15	(67)	0		47	(50)	33	(100)

other individuals. Detecting it may also be complicated by any tendency for vagrant individuals of rare species to be more likely to be recorded, because they are unusual.

(e) Resource usage

The explanation for the abundance–range size relationship which has attracted the most attention in recent years is that formulated by Brown (1984). He argued that for species which differ in only a few niche dimensions, if niches are multidimensional and spatial variation in the environment tends to be autocorrelated, there should be a positive correlation between abundance and range size. Species that can exploit a wide range of conditions locally and in so doing achieve high densities will also be able to survive in more places and hence over a larger area; the jack-of-all-trades is master of all. Those that can only exploit a narrow range of conditions will be unable to attain either high local densities or extensive distributions; the specialist is never very successful.

Brown's explanation of the relationship between abundances and range sizes assumes that more abundant and widespread species have broader environmental tolerances and/or an ability to use a broader range of resources. Indeed, Brown's (1984) line of argument has, in some quarters at least, become sufficiently accepted that often the distinction is not made between analyses of abundance–range size relationships and analyses of abundance–habitat breadth relationships. The evidence that this assumption is correct will be explored in some detail in Chapter 6. When various artefactual considerations are accounted for it is in fact rather poor.

The positive abundance–range size relationship may be explained by an alternative hypothesis equally based on resource usage. It is that those resources which are more abundant locally are also more widespread, and that those species that exploit resources which are locally abundant are as a result themselves able to become abundant and also widespread. The difference between those species which are common compared with those which are rare may thus lie in the abundance of the resources they require rather than in the variety of resources they are able to utilize. A review of the evidence for a relationship between species' abundances or range sizes and the availability of resources is postponed until Chapter 6. There is some support for such evidence.

Brown's (1984) hypothesis provides a plausible explanation for the exceptions to the positive abundance–range size relationship (Gaston and Lawton, 1990a). Negative relationships will occur when the habitat in which species' abundances are measured (the reference habitat) differs markedly from the spectrum of habitats or the most common habitat in the geographical region of interest. In this case, species which specialize in the reference habitat are unlikely to be successful in other areas. Thus, although they may reach high abundance levels in the reference habitat they will not become widespread. Conversely, those species which are able to exploit a variety of habitats may only be able to maintain relatively low abundances within the atypical reference habitat, although they are widespread. A relationship between abundance and range size would be absent when conditions fall somewhere between those producing positive and negative relationships. Gaston and Lawton (1990a) found that increasing the distinctiveness of the reference habitat, relative to the habitats in the region over which range sizes were measured, did indeed result in a shift of abundance–range size relationships from positive to negative. This result can, however, just as easily be explained in terms of resource abundance as in terms of the variety of resources.

Novotny (1991) suggests that whether positive or negative abundance–range size relationships are recovered does not depend on the distinctness or the abundance of the reference habitat. Rather, it depends on its ephemerality. Positive relationships, it is argued, are generated when studying ephemeral habitats, because these are dominated by species which are generalist and widespread. In contrast, negative relationships will be found for more permanent habitats, which are dominated by poorly distributed specialists. There seems to be some evidence for this position, but it hinges on the existence of a relationship between generalism and range size (see Chapter 6).

(f) Metapopulation dynamics

A positive relationship between abundance and range has been incorporated as an explicit assumption of some metapopulation models (e.g. the core–satellite model, Hanski, 1982a; Chapter 2). Metapopulation dynamics have also been used to explain the interaction. Two opposing viewpoints have been developed. Following Hanski (1991c) these can be termed the carrying capacity hypothesis and the rescue effect hypothesis.

The carrying capacity hypothesis (Nee *et al.*, 1991a) assumes that different species in an assemblage have different local carrying capacities, and that those which attain higher local population sizes have a lower extinction rate and/or a higher colonization rate than those which attain small local population sizes. It follows that the locally more abundant species will occupy more patches at equilibrium. Hanski (1991c) argues that if we assume that the local carrying capacity of a species is a reflection of its ecological specialism, then the carrying capacity hypothesis is Brown's (1984) hypothesis (see above) in another guise. There is, however, an important distinction. This genre of meta-population models assumes that all patches are equal, and hence that all species have the capacity to occupy all patches. Brown's (1984) hypothesis assumes that patches are very different and hence that not all species have the capacity to occupy all patches.

The rescue effect hypothesis (Hanski, 1991a) does not assume inter-specific differences in carrying capacities. Rather it assumes that immigration decreases the probability of a local population becoming extinct, and that the rate of immigration per patch increases as the proportion of patches which are occupied increases. Again a positive relationship between local abundance and number of occupied patches can result.

Attempts to test directly and distinguish between these models have, as yet, been limited (but see Hanski et al., 1993). They do, nonetheless, seem to make distinct and different predictions upon which such work could be based (Hanski, 1991c).

3.3.3 Synthesis

In summary, there are three groups of processes which might generate positive inter-specific abundance–range size relationships. They may result from: sampling artefacts, which is not particularly interesting; the way in which individuals are distributed, which places the interest in the reasons for aggregation; and/or other biological processes, of which resource usage and meta-population dynamics would have the most far-reaching consequences.

It seems unlikely that any one of these groups of processes has the power to explain entirely every observed abundance–range size relationship. Sampling considerations almost certainly contribute to all, and some general methodology for removing these effects is sorely needed. Some reliance has been put on the idea that the geographic ranges of the species of some taxa, especially birds, are sufficiently well documented in regions such as North America so that estimates of their relative sizes are reliable. While probably true of species' extents of occurrence this seems more doubtful for areas of occupancy.

Evidence for the various resource usage hypotheses is complex to untangle (Chapter 6) and judgement on their explanatory value must, for now, be withheld. Finally, if significant, metapopulation effects will be most important at meso-scales, leaving unexplained those patterns observed at smaller and larger scales.

3.4   CONCLUDING REMARKS

Many, even superficial, studies of rarity recognize some equality between those species which are rare in terms of their abundance and those which are rare in terms of their range size. The broad conclusion of this chapter is that this is a reasonable presumption – species which are rare on the basis of their abundance are often also rare on the basis of their range size. Data from a number of studies reveal a fairly constant proportion of species falling into the first quartiles of both abundance and range size frequency distributions (Table 3.3), though in some cases substantially less than the possible maximum. Having established this generalization, more work is now necessary to ascertain whether a theoretical base can be generated by which its detailed form can be predicted.

Species in an assemblage which are categorized as rare on the basis of abundance, categorized as rare on the basis of range size, and categorized as rare on the basis of both abundance and range size have been regarded by some as forming different types of rarity (Rabinowitz, 1981a; Rabinowitz et al., 1986; Arita et al., 1990; Kattan, 1992). Recognition of these different groupings tends to imply some commonality of the causes of rarity for the species in each category, though there is little reason to believe that this is in fact the case (Chapter 6). The different groups of species have been labelled in various ways. Those species which are widespread but never occur at high densities, that is those that are liable to be rare in terms of abundance but not in terms of range size, are often termed 'sparse' (e.g. Rabinowitz, 1981a). Rabinowitz (1981a) described them as '. . . those, which, when one wants to show the species to a visitor, one can never locate a specimen!'

Table 3.3 The number of species (*N*) included in different studies of the inter-specific abundance–range size relationship, and the number (*B*) and percentage (%) of these which fall into the lowest quartiles of both abundance and range size frequency distributions

	N	*B*	%
Bumblebees (Hanski, 1982b)	15	3	20.0
Birds (Bock and Ricklefs, 1983)	65	10	15.4
Zooplankton (Wright, 1983)	26	5	19.2
Birds (Bock, 1984)	70	10	14.3
Rodents (Brown, 1984)	12	2	16.7
Birds (Kouki and Häyrinen, 1991)	28	4	14.3
Ground beetles (Gardner, 1991)[a]	54	12	22.2

[a]As plotted in Figure 3.10.

4 Spatial dynamics

Nothing puzzles me more than time and space; and yet nothing troubles me less, as I never think about them.

C. Lamb (1810)

As defined in Chapter 1, the designation of species as rare or otherwise is made with reference to a particular area or spatial scale. Having been established, the categories might usefully be employed to investigate questions in other areas and at other scales. As emphasized, this approach agrees with the bulk of past applications of the term rare. Nonetheless, it still has to be determined whether there is any tendency for species which have been defined as rare in one area also to be defined as rare in another. That is, do species demonstrate spatially concordant rarity? In a similar vein, Schoener (1987, 1990) posed the question whether rarity (in his sense of occurrence in relatively few censuses and/or at relatively low abundances) was typically diffusive or typically suffusive. A species that was diffusively rare would be one which although rare in certain parts of its geographic range was common in others, while a suffusively rare species would be one which was rare everywhere.

Taking concordance as its central theme, this chapter explores the spatial dynamics of rarity. The early sections are concerned with approaches to determining the existence of concordance, the evidence for it, and consequences of spatial scale. Making the simplistic assumption that individual species are most likely to be viewed as rare (relative to other species) in those places where they are least abundant and/or least widespread, later sections concern the abundance structure of species' geographic ranges. The evidence for useful generalizations is examined, and possible differences in the structure of the ranges of rare and common species are explored.

4.1 FORMS OF SPATIAL CONCORDANCE

Spatial concordance in rarity can be tested for in several ways. Among the most obvious are to ask whether there is any significant tendency between different sites or areas for:

- individual species to fall consistently within the rare category;
- the species composition of the rare category to be the same;
- the rank abundances or range sizes of the rare species to be the same;
- the rank abundances or range sizes of all the species to be the same.

These alternatives will be termed 'numerical resolutions of concordance'.

They essentially form a nested hierarchy, in which the earlier criteria are likely to be true if the later ones hold. Concordance at each resolution can be tested for at various spatial scales.

Strictly, spatial concordance is tested by defining rarity at each site or in each region in terms of the set of species which occur there and belong to a larger assemblage defined at a spatial scale which embraces all the sites or regions in the analysis (the overall assemblage would probably be constrained on the basis of taxon or guild). However, much of the data potentially available to address the issue of concordance only enable contrasts to be made between the abundances or range size of the species in an assemblage at one site or region (effectively a reference assemblage) with the abundances or range sizes of those same species at other sites or in other regions. Provided few species occur at these other sites that might plausibly belong, on the basis perhaps of taxonomy or ecology (though obviously not in terms of location), to the assemblage at the reference site, then the results will be little different. If, however, there are many additional species, then this need not be so.

4.2   WHAT EVIDENCE?

It is commonplace to find published statements concerning individual species to the effect that either they tend to be rare everywhere that they occur, or they are rare in some places and common in others. One should remain wary of such pronouncements. Almost invariably they refer to the perceived or measured spatial variance in the abundances or range sizes of the particular species concerned, without reference to the abundances or range sizes of the other species with which they co-occur in different localities or areas. They are therefore only rare compared with some undefined standard, or compared with their own abundances or range sizes elsewhere.

In fact, it is remarkably difficult to find studies which have explicitly attempted to determine the spatial concordance of rarity, where rarity is defined relative to the measured abundances or range sizes of other species. This seems to reflect a general lack of interest in between-site comparisons of species rank (ordinal), rather than continuous, abundances or range sizes. While there have been numerous studies of the similarity of the faunas or floras of two or more sites, these have tended to follow one of two other courses. Some have measured similarity of species composition, using presence/absence similarity indices such as those of Sorensen, Jaccard and Preston (binary similarity coefficients; Southwood, 1978; Legendre and Legendre, 1983; Digby and Kempton, 1987; Krebs, 1989). Others have measured similarity of abundance structure, commonly using indices which reflect the distance between the faunas or floras in multidimensional space, where each axis of that space is the abundance of a different species (e.g. Euclidean distance measures; Legendre and Legendre, 1983; Digby and Kempton, 1987; Ludwig and Reynolds, 1988; Krebs, 1989).

Table 4.1 Comparisons of four artificial assemblages (each of five species) using ten common distance measures. The same species occur in all assemblages, and in three of them they share the same sequence of rank abundances; in the fourth this sequence is reversed. For each distance measure each comparison is scored from 1 for the two sites which were calculated to be most similar, to 6 for the sites which were calculated to be most different. Measures were calculated using the SUDIST program provided by Ludwig and Reynolds (1988), who provide details of their calculation. The measures are as follows (1) Euclidean distance measures: Euclidean Distance (ED), Squared Euclidean Distance (SED), Mean Euclidean Distance (MED), Absolute Distance (AD), Mean Absolute Distance (MAD), (2) Bray–Curtis dissimilarity index: Percent Dissimilarity (PD), and (3) relative Euclidean distance measures: Relative Euclidean Distance (RED), Relative Absolute Distance (RAD), Chord Distance (CRD), Geodesic Distance (GED)

Abundance matrix

Species	Sites			
	α	β	γ	δ
A	10	100	10	50
B	20	200	20	40
C	30	300	30	30
D	40	400	40	20
E	50	500	500	10

Sites		Distance measures									
		ED	SED	MED	AD	MAD	PD	RED	RAD	CRD	GDD
Abundances untransformed											
α	β	5	5	5	5 =	5 =	5 =	1	1	1	1
α	γ	2	2	2	2	2	3	4 =	4 =	2 =	2 =
α	δ	1	1	1	1	1	1	2 =	2 =	4 =	4 =
β	γ	4	4	4	4	4	2	4 =	4 =	2 =	2 =
β	δ	6	6	6	5 =	5 =	5 =	2 =	2 =	4 =	4 =
γ	δ	3	3	3	3	3	4	6	6	6	6
Abundances log10 transformed											
α	β	5	5	5	5 =	5 =	5 =	1	1	1	1
α	γ	1	1 =	1	1	1	1	2	2	2	2
α	δ	2	1 =	2	2	2	2	5	5	5	5
β	γ	4	3 =	4	4	4	3 =	4	3	3 =	3 =
β	δ	6	6	6	5 =	5 =	5 =	3	4	3 =	3 =
γ	δ	3	3 =	3	3	3	3 =	6	6	6	6

Methods of determining the concordance of rarity lie somewhere between these two approaches. Determination of whether a given species consistently falls within the rare category requires some information on species' abundances to determine the limits to this category. Likewise, while those techniques for determining the compositional similarity of faunas and floras can be applied only to the rare category, abundance information is necessary for its delineation. Conversely, ascertaining whether the rank abundances of species in the rare category, or in an entire assemblage, are more constant than would be expected by chance, does not necessitate as much information as for measuring the similarity of the abundance structure of faunas with distance measures. Perhaps more importantly, there is no simple relationship between the similarity of two assemblages based on a distance measure and the similarity of the rank abundances of their species (measured using, say, a rank correlation such as Kendall's tau; Siegel, 1956). The extent to which similarity of rank abundance influences the relative magnitude of distance measures depends on the particular distance measure used and whether abundance data are transformed (Table 4.1); the value of indices is also dependent on sample size and diversity (Wolda, 1981).

The absence of adequate published examples by which the spatial concordance of rarity can be judged probably does not pose a major difficulty in surmising likely patterns. Concentrating on the higher of the resolutions of concordance listed earlier, that of constancy in the rank abundances of an entire assemblage, high concordance is more probable when:

- Spatial variances in the abundances or range sizes of individual species are low.
- The spread of the mean abundances or range sizes of species in an assemblage is broad relative to the variance about each mean.
- The species' composition of the assemblages being compared is similar.

The degree to which these conditions are met is likely to be a function of three, related, things: how closely the habitats of the sites being compared resemble one another; how widely separated the sites are in space; and the spatial scale of each site. The more dissimilar habitats are and the more distant sites are from one another, the less likely it is that high concordance will be recovered. The effects of spatial scale are potentially quite complicated and will be dealt with in the following section.

By studying the relative abundances of plant species in chalk grassland, Mitchley and Grubb (1986) reduced habitat differences between the assemblages they compared. They found significant concordance in the abundance ranks of species, for: samples from different parts of the same sites; samples from two sites approximately 150 km apart; literature-based comparisons of several different sites across Britain.

While constancy, across different sites, of rank abundances or range sizes across all the species in an assemblage can broadly be regarded as evidence for concordant rarity, some care must be exercised. Common and rare species may not contribute equally to measures of concordance. In particular, common species may contribute disproportionately.

As indicated, constancy of the ranked abundances or range sizes of all the species in an assemblage is one of the stricter tests of spatial concordance. It is highly probable that the necessary conditions for less restrictive tests will be met in many instances where stricter conditions are not. These other tests are easier to perform.

4.3 SPATIAL SCALES

Regardless of the precise criterion, the frequency with which rarity is observed to be spatially concordant may depend crucially upon the spatial scale at which it is examined. At micro-scales, most species that are found to be rare are likely to be substantially more abundant or widespread elsewhere and are likely to lie outside the rare category in many of these places, and hence rarity is likely to be predominantly diffusive. This was Raunkiaer's explanation for bimodal frequency distributions of species' micro-ranges: the peak of restricted species comprised those adapted to and more common in other habitats, while the peak of widespread species comprised those adapted to the habitat of the study region (Chapter 2). Terborgh *et al.* (1990) report that the largest proportion of the bird species categorized as rare on their 97-ha floodplain forest plot were more abundant in other nearby habitats, the implication being that they would not be in the rare category in these places. Dressler (1982) says of orchids that ' "rare" usually means "hard to find where we looked". Again and again, "rare" species prove to be frequent in the right habitat. . . '. Stevens' (1989) hypothesis that mass effects are largely responsible for high local species' richness in lowland tropical forests suggests that the proportional contribution of diffusive rarity to local richness may be habitat dependent.

At intermediate (meso-) scales, rarity is likely to be more of a mix of the diffusive and suffusive. Some species which are rare in a given area are probably rare throughout many other areas, while some at least are not. Thus, Owen and Gilbert (1989) found that the meso-ranges of hoverfly species in each of several regions of Europe were positively correlated. Likewise, Jablonski and Valentine (1990) found significant, though weak, correlations between the sizes of the ranges of bivalve species within a faunistic province and outside it.

The proportion of species which are diffusively rare at meso-scales may be influenced particularly by the way in which vagrants are treated, because they contribute most to levels of species' richness at those scales (Chapter 1). If regarded as part of the species assemblage, they will tend to increase the diffusive component, because they are almost by definition more abundant and widespread elsewhere.

Perhaps the most interesting question about the concordance of rarity is to ask over what proportion of a species' geographic range it is rare at meso-scales. As others have found, quantitative data with which to address this issue are remarkably elusive (e.g. Schoener, 1987, 1990; Samways, 1990). Samways (1990) suggests that there is circumstantial evidence to indicate that in the epigaeic ant assemblages of South Africa rarity is suffusive. Indeed, for most groups of organisms it seems possible subjectively to identify individual species which are, relative to a given assemblage, consistently rare across the

Table 4.2 Studies of inter-specific relationships between range sizes when one area is nested inside another. All relationships were reported to be positive with the exception of those with asterisks

Source	Taxon	Comparison
Hengeveld and Hogeweg (1979)	Beetles	The Netherlands *v.* Europe
Hodgson (1986b)	Plants	Sheffield region *v.* Britain
Rapoport *et al.* (1986)	Plants	Berkshire *v.* Britain
Williams (1988)	Bumblebees	Kent *v.* Britain
Owen and Gilbert (1989)	Hoverflies	Co. Laois *v.* Ireland
Ford (1990)	Birds	New England *v.* Australia
Gaston and Lawton (1990b)	Fish	UK *v.* Europe
Jablonski and Valentine (1990)	Bivalves	Oregonian province *v.* Global
Buzas and Culver (1991)	Foraminiferans	Within and between regions for five geographic regions around N. America
Daniels *et al.* (1991)	Birds	Oriental region *v.* Global Indian subregion *v.* Oriental region Malabar province *v.* Indian subregion Indian subregion *v.* Global* Malabar province *v.* Global Malabar province *v.* Oriental region
Lahti *et al.* (1991)	Plants	Finland *v.* Northern hemisphere
P.H. Williams (1991)	Bumblebees	Kashmir *v.* Global*

whole of their geographic ranges at meso-scales. To what extent this is a matter of perception remains unclear.

The concept of spatial concordance of rarity is difficult to apply at macro-scales. Those species which are rare will by definition often simply not occur elsewhere.

4.4 NESTED AREAS

Consideration of the effects of spatial scale on the probability of observing concordance raises the question of how likely it is that a species that is rare in a given area will be rare across a larger area within which the first is nested. There have been a number of analyses of the relationships between the abundances or range sizes of species at different spatial scales, in which the smaller areas lie within the confines of the larger area (Table 4.2, Figures 4.1 and 4.2).

Such analyses evidently bear a very close relationship to those of inter-specific abundance–range size relationships (Chapter 3), especially to those of the latter which measure abundances in only a single area or several areas contained within a region somewhat less than the size of the range.

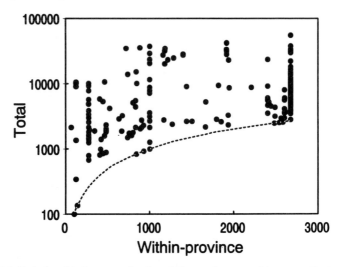

Figure 4.1 Relationship between the size of the total geographic ranges (km) of marine bivalve species and the size of their ranges within the Oregonian province of the northeastern Pacific shelf (km). The dotted line indicates where the within-province range is the total geographic range. Note that one axis is log-transformed. (From data in Jablonski and Valentine, 1990.)

In almost all cases the data sets have comprised only those species found at the smaller scale. Repeatedly it has been found that there is a significant tendency for those species which are most abundant or most widespread at small spatial scales to be most abundant or widespread at large spatial scales (Table 4.2; Collins and Glenn, 1990).

In one of the only studies to consider multiple nested spatial scales, Bock (1987) compared the abundances of 62 species of Arizona landbirds at four different scales. The species that were most abundant on study plots and among habitat types in the Huachuca mountains were also most abundant across Arizona and throughout the western USA. These data also reveal that the species which were recorded from the most plots locally also tended to be more widely distributed at larger spatial scales. In both cases, the strength of the correlations tends to decline with the increasing disparity between the scales.

A relationship between local and regional range size has also been observed in the altitudinal dimension. The elevational amplitudes of Andean birds on a transect in the Cordillera Vilcabamba were related to their elevational amplitudes determined across their entire geographic ranges (Graves, 1988).

In most instances where abundances or range sizes at different scales have been examined their relationships have been comparatively weak. There is also a concern that they often have an artefactual component, in that abundances (if absolute) and range sizes in the larger region will inevitably be equal to or greater than those in the smaller (Jablonski and Valentine, 1990). This generates a lower bound to the relationship which increases with the magnitude of

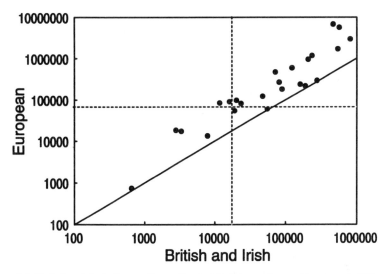

Figure 4.2 Relationship between the estimated total numbers of pairs of different seabird species in Europe and in Britain and Ireland. Solid line is where the entire European population is in Britain and Ireland. (From data in Lloyd *et al.*, 1991.)

the abundance or range size in the larger region, and means that the correct null hypothesis is not a slope or correlation of zero. Support for the importance of this artefact comes from the observation that there are instances in which the relationship takes on something of a triangular form, with species which are rare at the smaller spatial scale having abundances or range sizes of a variety of magnitudes at the larger scale, while species which are common at the smaller scale are always abundant or widespread at the larger (Figure 4.1; Hengeveld and Hogeweg, 1979).

Whether artefacts can entirely explain the relationship between species' abundances or range sizes at different scales is unclear. However, given the weakness of many documented examples, suggestions that such relationships may be used as a practical basis for inferring the relative sizes of the geographic ranges of species on the basis of their range in a smaller area (Jablonski and Valentine, 1990) seem premature. Various candidates for biological explanations of relationships between species' abundances or range sizes measured at different spatial scales can be found by slight modifications of some of those offered for inter-specific abundance–range size relationships (Chapter 3). Thus, they could result if, at both scales, appropriate habitat for rare species tended to be more limited than that for common species, or if rare species had consistently narrower environmental tolerances.

Collins and Glenn (1990) interpret significant positive correlations between species' abundances at two spatial scales (in their case, the average cover of grassland plant species in each of two small blocks (10 m × 10 m) and their average cover at the regional level) as a rough measure of the self-similarity of patterns of abundance (i.e. community structure is essentially fractal, in that the larger unit is composed of numerous smaller units similar in structure to the larger unit).

4.5   ABUNDANCE STRUCTURE OF GEOGRAPHIC RANGES

4.5.1 General patterns

Although it need not be so simple, individual species are most likely to be regarded as rare in relation to other species in those places where they are either least abundant and/or least widespread. Are there any generalizations that can be made as to the parts of their geographic ranges where these conditions are most likely to pertain?

It has long been held that, in general, the abundance of a species peaks at the centre of its geographic range and diminishes towards the margins (see Hengeveld and Haeck, 1981 for references). A variety of additional constraints upon this basic idea have been suggested, to the effect that the spatial distribution of abundances tends to be unimodal, that the decline in abundance from range centre to limits tends to be monotonic, or that the structure of a species' range can typically be regarded as Gaussian (densities along a transect running between the range margins and through the range centre fall along a normal curve).

Empirical data upon which to base analyses of the abundance structure of geographic ranges are limited. Nonetheless, a number of attempts have been made to ascertain patterns for one or more species (e.g. McClure and Price, 1976; Bock et al., 1977; Taylor and Taylor, 1979; Hengeveld and Haeck, 1981, 1982; Brown, 1984; Carter and Prince, 1985; Emlen et al., 1986; Root, 1988; Cammell et al., 1989; Wiens, 1989; Roberts et al., 1992; Svensson, 1992). The most data are available for birds. Root (1988), for example, has mapped the winter abundances of more than 200 North American species. Other studies concern considerably fewer species, such as Svensson's (1992) examination of the abundance patterns of three species of *Gyrinus* beetles in northern Europe. While most studies concern contemporary distributions, a historical dimension can be added from such sources as isopoll maps (e.g. Huntley and Birks, 1983; Huntley and Webb, 1989).

It is difficult to derive anything beyond the broadest of conclusions from this work. There are two primary reasons. First, there is seldom any attempt to determine the fit of statistical models to data on the distribution of abundances across species' geographic ranges. Most interpretations of the patterns are based on visual impressions. While not of themselves entirely unhelpful, such approaches are of limited value when debating the usefulness of given generalizations. What may appear to some as convincing examples of particular broad patterns, may appear to others to provide little agreement with any simple models. Thus, for example, while few supporters of the Gaussian model would dispute that there are plainly examples for which it does not provide an adequate fit, it is seldom made clear whether the frequency of these exceptions is sufficient to invalidate the model or not. Differentiation between particular models will be made more difficult because data on the spatial abundances of species are more often available for a small portion of their geographic range than for sites or areas spread across the whole of it. Ideally, the abundance structure of geographic ranges needs to be analysed using techniques of spatial autocorrelation applied to abundance data obtained across much of a species' range.

Second, patterns in the spatial distribution of abundances are strongly influenced by the resolution and methods of mapping (Wiens, 1989). Thus, models which appear to provide good descriptors of a species' range structure for one resolution and methodology may not do so for another. Different studies map the abundances of species in different ways and their results can only be compared with caution. Indeed, it is obvious that no species is continuously distributed in space, so that when it is mapped at a fine enough scale there will be no smooth change in abundances from one point to another.

The complexities of spatial scale are further complicated by the impossibility of mapping species abundances instantaneously. Different mapping schemes are seldom based on records summed over equivalent periods. It can be argued that what is primarily desired is a model that captures the broad form of the abundance structure of ranges, when appropriately averaged over space and time. It remains to be determined what is appropriate averaging.

Examples of many different spatial distributions of species' abundances exist, these include instances of unimodal and multimodal peaks, of sharp truncations and of gradual declines (Figures 4.3–4.5). Although it is simply to perpetuate subjective judgements, I suggest that while a Gaussian model may for some purposes provide a useful way of viewing species ranges, it has little generality (Carter and Prince, 1988; Wiens, 1989). It seems reasonable to maintain that the abundances of most species tend to fade out toward range edges rather than to drop abruptly, except where there is a geographical discontinuity. However, although abundances at range limits are commonly lower than those in many parts of a species' range, they are often no lower than those of areas well within the range limits. There is, for example, no evidence of a decline in population size towards the northern limits in Britain of prickly lettuce (*Lactuca serriola*), indeed some of the large populations occur at the limits (Carter and Prince, 1985). Relatively high abundances at range limits may be associated with species whose ranges are expanding (e.g. Chiang, 1961).

We might expect from the previous chapter that levels of abundance and occupancy will tend to change in parallel across individual species' geographic ranges. Svensson (1992) provides some evidence for such a pattern for species of *Gyrinus*. Conversely, Carter and Prince (1988) argue that while in general there is evidence for a decline in the frequency of populations (occupancy) towards range limits, there is little evidence to support the view that populations become smaller. Notwithstanding, similar conclusions can be reached regarding the distribution of levels of occupancy across species' geographic ranges, as are reached for their abundances. Brussard's (1984) examination of

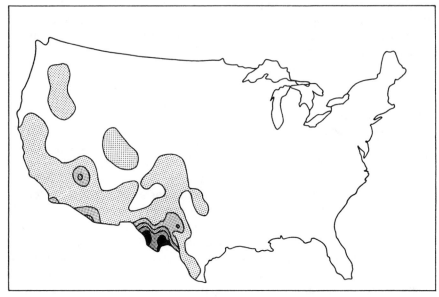

Fig. 4.3 (a)

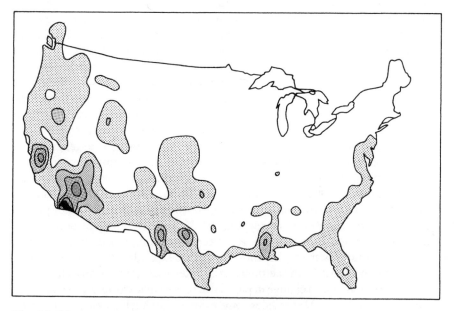

Fig. 4.3 (b)

Figure 4.3 The abundance structure of the geographic ranges of three species of wrens (Troglodytidae). (a) Rock wren (*Salpinctes obsoletus*); (b) marsh wren (*Cistothorus palustris*); and (c) Carolina wren (*Thryothorus indovicianus*). Abundances are scaled to the maximum abundance in each case, with the edge of the range defined as 0.5% of this value, and the contour intervals at 20%, 40%, 60% and 80%. (Modified from Root, 1988.)

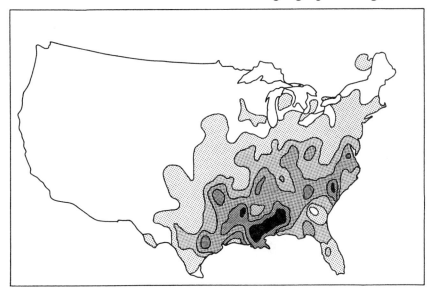

Fig. 4.3 (c)

the range maps of 200 species of North American forest trees revealed that in about a third of cases there was a central, essentially continuously inhabited area surrounded by more patchily occupied peripheral areas, in another third there was more than one distributional centre, and the rest were distributed in dendritic patterns, as islands or archipelagoes, or as highly disjunct isolates.

Hengeveld and Haeck (1982) examined changes in levels of occupancy across species' geographic ranges for a number of data sets in a less direct fashion. They determined whether there was a relationship between the frequency of occupancy of a species in an area and the position of that area relative to the species' entire geographic range (marginal, submarginal, subcentral or central). The conclusion was that their analyses supported a model in which frequencies of occurrence were highest at the centre of geographic ranges and lowest at the periphery – 'The geographical distribution of species' abundance is as a rule heterogenous, with the highest abundances occurring near the centres of the species' ranges and the lowest at the margins . . . we feel that this trend may be considered as a general biogeographical rule.' This sweeping statement rests on the assumption of a strong inter-specific relationship between regional density and frequency of occurrence in each of their data sets, and there being no relationship between range category and total geographic range size or between total geographic range size and frequency of occurrence. The former is possible; the latter may be less likely.

In summary, it seems doubtful whether empirically any but the vaguest of generalizations can be made as to the parts of their geographic ranges where, because their abundances or their levels of occupancy are lowest in those regions, individual species have the greatest probability of being rare with respect to other species.

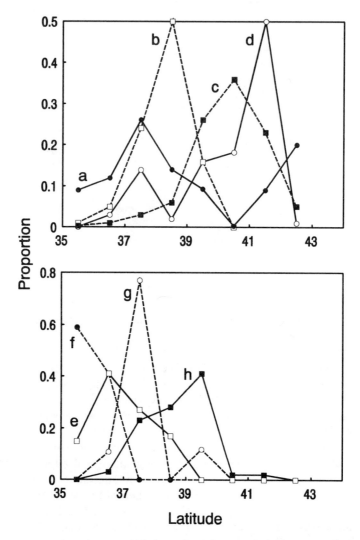

Figure 4.4 Proportion of the total number of individuals of each of eight *Erythroneura* leafhopper species (on the American sycamore, *Platanus occidentalis*) collected from eight zones along a latitudinal gradient. *Erythroneura* species are (a) *arta*, (b) *usitata*, (c) *lawsoni*, (d) *torella*, (e) *bella*, (f) *hymettana*, (g) *ingrata* and (h) *morgani*. (Redrawn from McClure and Price, 1976.)

4.5.2 Geographic range structure and rarity

Some sense may perhaps be made of the appearance of idiosyncrasy in the abundance structure of species' geographic ranges by rigorous attempts to look for differences in these structures for the ranges of different groups of species. One might perhaps hypothesize that plant species whose occurrence is limited by geology might have ranges of rather different abundance structure to those whose occurrence is limited by herbivory. Geology is more likely to change abruptly in space than is herbivory.

One obvious inter-specific difference on which to look for differences is that of rarity itself. Several differences in the abundance structure of the geographic ranges of widely and narrowly distributed species have been postulated:

(i)   Of the potential differences, perhaps the most often cited is that rare species have more fragmented geographic ranges than common species (e.g. Griggs, 1940). A similar claim has been made with regard to many tropical species when compared with their temperate counterparts (Diamond, 1980), which may amount to much the same thing if rarity is defined at a global level. Arguments as to the comparative fragmentation of the ranges of common and rare species can verge on the circular. The ranges of rare species will inevitably tend to be more fragmented, simply because they occupy fewer sites. The important question is whether rare species have ranges that are more fragmented than one would expect. Analyses founded on fractal descriptions of species' geographic ranges (Chapter 3) may provide the best way to study this in detail.

Quinn *et al.* (1994) have explored the range structure of some scarce plant species in Britain (part of their geographic ranges) and found them to be more aggregated in their patterns of occupancy than expected by chance (see also Cornell, 1982). They also find various correlates of this degree of aggregation, including dispersal ability (poor dispersers are more strongly aggregated).

(ii)   Bowers (1988) documents a negative correlation between the sizes of the geographic ranges of species of desert heteromyid rodents and the spatial variability of their abundances at sites in local regions. The abundances of rare species are less evenly distributed than are those of common species. The question of the evenness of the spatial distribution of abundances for species of differing mean local abundance (both within small regions, as in Bowers' analysis, and across larger areas) or geographic range size is an important one. It is severely complicated by difficulties in inter-specific comparisons of variability measures, which make Bowers' findings hard to interpret (see Chapter 5 for a list of these problems as they pertain to temporal variabilities; McArdle *et al.*, 1990). However, it seems safest to assume that rare species have the spatially more variable abundances.

Bowers argues that for species to have a comparatively even distribution of abundances in space they must have the capacity to use a wide range of resources, and therefore that his findings are consistent with Brown's (1984) theory of abundance and range size (Chapter 3). I see no obvious justification for the initial premise, and the full argument necessitates that abundance and niche breadth be directly related (see Chapter 6).

If indeed the local abundances of rare species are more spatially variable than are those of common species this raises some potentially interesting issues about the effects of the high risk of extinction of small populations (Chapter 5) on the abundance structure of geographic ranges. One might perhaps have imagined that the low mean local abundances of rare species would have

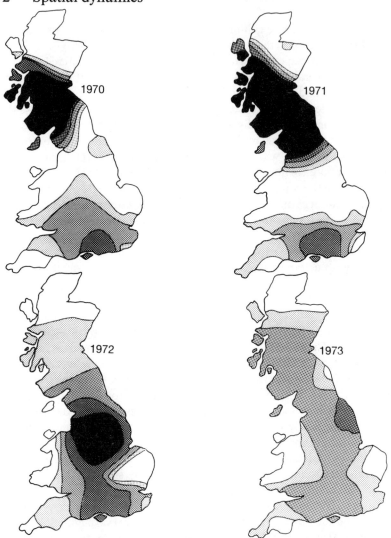

Figure 4.5 The changes in the density (on a logarithmic scale) of the elder aphid (*Aphis sambuci*) across Britain between 1970 and 1977. The winter and summer hosts of the species are distributed throughout Britain. (Redrawn from Taylor and Taylor, 1979.)

resulted in their having low spatial variability in these abundances, because populations at levels below the mean would have a high risk of extinction.

(iii)   The possibility has been raised for Australian birds, that species with large geographic ranges tend to occur with low frequencies, relative to their maximal frequency of occurrence, over a greater proportion of their ranges than do species with small ranges (Schoener, 1990). No explanation has been offered for this pattern, and it remains to be seen whether it generalizes or not.

Figure 4.5 continued

4.6   CONCLUDING REMARKS

A synthesis of what is known of the spatial dynamics of rarity comprises more a collection of impressionist than of still-life sketches. Impressions are, almost by definition, subjective. There is no need, however, for such a state of affairs to persist. A number of schemes have collected, and in many cases continue to collect, information on the abundances of species at multiple sites scattered over large geographic areas (e.g. Root, 1988; Marchant *et al.*, 1990). These data could provide the basis for comprehensive analyses of patterns in the concordance of rarity and in the abundance structure of ranges.

5 Temporal dynamics

'The balance of nature' does not exist and perhaps never has existed.

C. Elton (1930)

Why do the numbers of every kind of living thing stay so roughly constant? Why do the common stay common and the rare stay rare?

P. Colinvaux (1980)

Questions about the effects of spatial scale on which species are recognized as being rare and about other features of the spatial dynamics of rarity (Chapter 4) find parallels in the temporal dimension.

In this chapter, various facets of the temporal dynamics of rarity are explored. As in the previous chapter, I begin with the issue of concordance. For convenience, several, essentially arbitrary, combinations of spatial and temporal scales are considered separately. At the limit, of course, no species can remain rare indefinitely. This leads to the relationship between rarity and temporal persistence, which provides the subject of the second major section of the chapter. Here, persistence in geological time and in ecological time are distinguished. Regardless of generalizations about the temporal dynamics of rarity, many rare species are plainly on a trajectory to extinction, and the third section addresses the forms this trajectory might take. Finally, a fourth and brief section highlights some ideas about the relationship between rarity and ecological success.

5.1 TEMPORAL CONCORDANCE

Perhaps the most important question we can ask of the temporal dimension to rarity is whether those species which presently we regard as rare have also been so in the past and are likely to be so in the future. That is, are species suffusively or diffusively rare through time (rare throughout a time period, or rare at some times but not at others).

As in considering spatial concordance, the temporal concordance of rarity can be examined at a variety of numerical resolutions. Spatial concordance can be studied over very short periods of time, the brevity of which is limited only by practical considerations, thereby substantially reducing any complications of temporal scale. It is not, however, helpful to minimize the spatial scale in a similar way when considering temporal concordance!

The breadth of spatial and temporal scales at which temporal concordance can in principle be investigated encompasses a huge variety of abiotic and

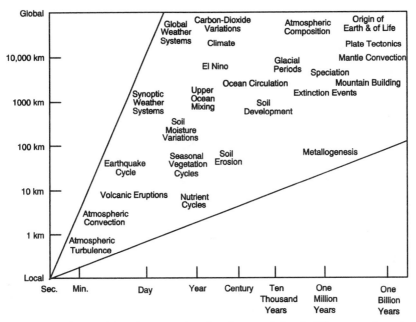

Figure 5.1 The spatial and temporal scales characteristic of different earth system processes. (Redrawn from National Aeronautics and Space Administration, 1988.)

biotic processes (Figure 5.1). In practice, sufficient data are available to draw robust conclusions about concordance at only a few of these combinations.

5.1.1 Local abundance

Several of the possible numerical resolutions of the concordance of rarity have been investigated with respect to the temporal dynamics of species' local abundances, in the context of community persistence. Rahel (1990) recognized three numerical resolutions of community persistence, the presence or absence of species, their relative (or rank) abundances, and their absolute abundances (Figure 5.2; use here of the terms 'relative' and 'absolute' needs to be distinguished carefully from that of section 2.1.3). These overlap with some of the higher numerical resolutions of the concordance of rarity listed in the previous chapter. Stability at either of the last two of Rahel's numerical resolutions would suggest that rarity was temporally concordant.

In analysing data for a variety of assemblages, Rahel (1990) concluded that most were stable in terms both of species' presence or absence and of the abundance rankings of component species. Most assemblages were unstable in terms of the absolute abundances of component species. These results were true, for example, at each of three sites for shrub-steppe birds in the western USA, censused annually from 1977 to 1983. Indeed, it seems that in general, rather few local assemblages are temporally highly unpredictable in species

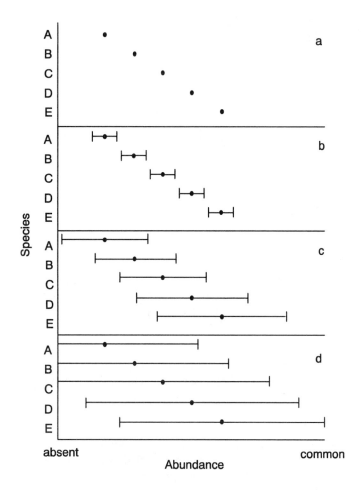

Figure 5.2 Patterns of temporal stability of a hypothetical assemblage of five species. (a) No variation in species' abundances; (b) species' abundances vary but abundance rankings remain constant; (c) species' abundances show sufficient variation that abundance rankings are not stable; (d) species' abundances show sufficient variation that neither abundance rankings nor species' presence or absence are predictable. (Redrawn from Rahel, 1990.)

composition and rank abundance (Lawton and Gaston, 1989; see also Gilbert, 1991; Crowley and Johnson, 1992; Obeso, 1992). This stability can be demonstrated through experimental manipulation (see Meffe and Sheldon, 1990 for references). Thus, for example, Meffe and Sheldon (1990) analysed the structure of fish assemblages at 37 stream sites, before and nearly one year after they were experimentally defaunated. For all sites combined, both the rankings of species' abundances and species' absolute abundances, before

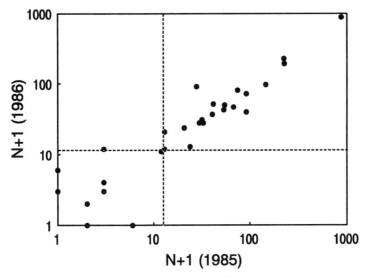

Figure 5.3 Relationship between the total number of individuals (+1) of various fish species collected from 37 South Carolina stream sites before (1985) and nearly a year after (1986) they were experimentally defaunated. (From data in Meffe and Sheldon, 1990.)

andafterdefaunationweresignificantlypositivelycorrelated(Figure 5.3). The species' richness and summed densities at each site did not differ significantly between the pre- and post-defaunation samples, and were correlated. The defaunation did not significantly change the structure of the local assemblages, nor the densities of most species when paired by sites. In short, the structure of these assemblages proved to be very resilient.

Temporal stability in the structure of an assemblage results from the existence of bounds to the local abundances of each, or most, of the component species. Classically, these bounds would be visualized as tight constraints resulting either from resource limitation, or from the action of predators, parasites and/or competitors maintaining populations well below carrying capacity. Indeed, there are systems where this seems to be the case (e.g. some raptor assemblages). For significant concordance, abundances need, however, only to be tightly bound relative to the spread of the mean abundances of the different species. The abundances of individual species may vary over orders of magnitude without necessarily violating this condition (by analogy, although the frequency distribution of the local abundances of a species in space is strongly right-skewed, it may, relative to the other species present, still be rare in most places where it occurs). This means that concordance may result from substantially less deterministic processes than those classically proposed (Ebeling *et al.*, 1990). These might include the immigration/emigration of individuals from a regional pool. Indeed, while it is possible to envisage various 'ideal' types of communities which are structured by different processes (Strong *et al.*, 1984; Lawton and Gaston, 1989), levels of concordance will often not provide insights into how well real assemblages conform to them (Ebeling *et al.*, 1990 and references therein).

Although in general, at the scale of a local assemblage, rarity tends to be temporally concordant, some caveats are necessary.

- Despite seeming to be in the minority, local assemblages have been documented that are unstable either in composition or in rank abundances. Townsend *et al.* (1987), for instance, found marked changes in both the composition and ranking of species at stream sites in surveys separated by 8 years. Examples of systems which do not show resilience when experimentally perturbed also exist (e.g. Sale, 1988 and references therein).

- Significant temporal concordance in the species' composition of a local assemblage may not mean that that composition remains identical. Moreover, those changes which do take place are most likely to be among the rare species. These tend to have a greater probability of local extinction (see below). Such turnover may be substantial (Williamson, 1989a). Rogers (1983) observed no significant between-year differences in the floristic composition and frequency (proportion of plots in which a species occurs) of forest herb assemblages before and after a severe natural perturbation. Those floristic differences which did occur were largely restricted to rare species and could potentially be explained by the large random errors associated with analyses of these species.

- In a similar vein, significant temporal concordance in species' rank abundances need not imply that the stability of those ranks is even throughout an assemblage. A common observation is that much of the significance derives from stability in the ranks of the most abundant species. There may be some substantial shifting of abundance ranks among the rare species, although they continue to remain rare.

- The degree of temporal concordance observed may differ between habitats. Thus, it is stronger for fish assemblages in pools than in riffles (Angermeier and Schlosser, 1989). Stability is likely to depend on the frequency of disturbances and the rapidity of successional processes.

- Temporal concordance may depend on the life histories of the taxa concerned. Floras dominated by annuals are expected to be more susceptible to perturbations than those dominated by long-lived perennials (Rogers, 1983).

- Concordance in the composition and rank abundances of an entire assemblage is one of the higher numerical resolutions at which the concordance of rarity in local abundances can be examined. As when examining the spatial concordance of rarity, it is difficult to find studies of temporal concordance at coarser numerical resolutions. However, it remains likely that these conditions will be met with far higher frequency.

- Whether a given assemblage appears to be temporally stable or unstable may depend both on the spatial and the temporal scale (Davis and Van-Blaricom, 1978). An assemblage which is stable at lower scales may be unstable at larger scales, and *vice versa*.

5.1.2 Range sizes and large spatial scales

One consequence of any tendency for the species' composition of individual sites to remain approximately constant through time is likely to be temporal concordance in species' meso- and macro-range sizes. Individual species will consistently occur at roughly the same number of sites, and those species which are rare by virtue of occupying few sites will remain rare. In fact, the conditions for temporal concordance in species' range sizes are rather less stringent than this. Species need not occupy the same sites from one time period to the next in order for the ranking of their range sizes to remain approximately constant, provided the numbers of sites at which each species becomes extinct roughly equals the number it colonizes. Indeed, they need not even occupy the same numbers of sites, only have sufficient bounds on the numbers they do occupy (as for abundances).

The general constancy of species' relative range sizes is difficult to judge. Discounting for a moment the effects of human activities, the geographic range sizes of many species seem to be reasonably stable over time scales of many years. Indeed, in areas which have remained pristine, it is frequently possible to find the same species present as were recorded in earlier decades or centuries, often at much the same densities. Felton (1974) cites several examples of the apparent persistence of populations of aculeates (Hymenoptera). These include the extreme case of the small sphecid *Crossocerus exiguus*, of which only three specimens were, at the time of writing, known from Britain. These were collected in 1896, 1935 and 1967, all from localities about 20 km apart (since then the species has been found at a few additional sites in south-east England; Falk, 1991).

Of course, the likelihood of encountering the same species at the same sites at different times may depend on the mobility and longevity of its individuals. Poor dispersal abilities and long life spans will enhance this probability. The majority of documented cases of localities being revisited after many years and the continued presence of a species noted, concern rare species (particularly plants). Interestingly, it has been claimed that both dispersal ability and longevity are correlates of range size and of abundance, such that rare species have poorer dispersal abilities and longer life spans (Chapter 6). However, no direct link between these sets of observations and any apparent differential persistence of rare species should be made. The continued presence of a rare species at a site is more likely to be regarded as noteworthy than is the continued presence of a common species, and there are grounds for believing that populations of rare species are less likely to persist than are those of common species (see below).

Any strong tendency for species to persist at sites weakens the validity of much of classical metapopulation theory (Levins, 1969; Gilpin and Hanski, 1991). This regards the recurrent local extinction of populations as a component of the natural dynamics of species' regional occurrences. A recent review confirms the low frequency of documented cases of classical metapopulations, although how representative a sample these constitute is not known (Harrison, 1991).

Against any picture of stability in relative range sizes must be placed the undoubted evidence of instability in the sizes of the ranges of some individual species. Steady changes in the patterns of occurrence of species (range displacement, contraction or expansion) have been documented as responses to climatic trends over periods of a few decades, as well as resulting from the invasion of new predators, diseases or competitors. The white admiral butterfly (*Ladoga camilla*), for example, underwent a striking expansion of range in England in the 1930s and early 1940s, which was associated with favourable weather conditions (Pollard, 1979). Diamond (1984a) cites many instances of climatic changes in species' occurrences. Of course, the question often remains whether these changes themselves result in changes in the ranking of species' range sizes.

Frustratingly, direct comparisons of the past and present range sizes of species can seldom be made from published maps. While these often distinguish between localities which were occupied in the past but are not now and those which are presently occupied, they often do not distinguish between localities which are occupied now but were not in the past. Nor do they indicate in detail how levels of sampling effort have changed. Recent analyses of data on the areas of occupancy of butterfly and of carabid species in the Netherlands have sought to overcome both of these problems with reference to the raw data from which maps have been constructed (Turin and den Boer, 1988; van Swaay, 1990). In these studies the authors have sought to discriminate groups of species, the range sizes of which have changed in different ways (e.g. stable, increased, decreased) over the last century or so. Reanalysis of Turin and den Boer's (1988) data reveals a broad correlation between the areas of occupancy of the 80 species analysed (which exclude any which occur only locally) in different decades (Figure 5.4). A pattern of increasing correlation with decreasing temporal separation of decades is obscured by the results pertaining to a single decade (1890–1899).

Even major habitat destruction has left substantial similarity in the relative macro-range sizes of some South-East Asian primate species (Figure 5.5).

On balance, it seems likely that range sizes tend to show temporal concordance over the time scales of years to decades.

5.1.3 Longer time scales

Uncertainties about the levels of concordance of rarity (in terms of abundance or range size) over comparatively short time scales can be resolved by direct observation. Over longer time scales (thousands to millions of years) this is not, of course, possible. Rather, various indirect fragments of evidence (preferably divorced of their attendant errors) have to be pieced together.

Central to an understanding of the long-term concordance of rarity is the issue of the long-term temporal dynamics of the sizes of species' geographic ranges. The greater the stability of species' range sizes the more likely rarity is to be concordant. A high degree of inconstancy in range sizes is unlikely to result in concordance unless changes in range size are approximately synchron-

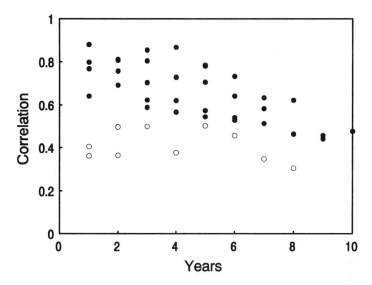

Figure 5.4 Spearman rank correlation coefficients between the range sizes in different decades of 80 species of carabid beetles in the Netherlands, plotted against the numbers of decades separating the two being compared. Open circles are for the period 1890–1899. (From data in Turin and den Boer, 1988.)

ous for all the species in an assemblage. There is no consensus as to the stability of species' range sizes over long periods; instead it is easier to recognize two divergent viewpoints which respectively emphasize dynamism and stasis in range sizes.

The case for dynamism in species' range sizes is largely based on the observation that long-term changes in climate result in temporal changes in the abundance and spatial occurrence of species, and hence in the structure and composition of local and regional assemblages. The best information on such responses has been generated from work on pollen data, particularly over the period since the end of the last glacial (10 000 year BP). This has emphasized the individualistic response of species to environmental change, and the transience of assemblage composition (Davis, 1976; Huntley and Birks, 1983; Webb, 1987; Hunter *et al.*, 1988; Huntley, 1988, 1991; Huntley and Webb, 1989).

Dynamism also tends to follow from Van Valen's (1973a) 'Red Queen' hypothesis (Ricklefs and Latham, 1992). This postulates that evolution of the other organisms with which a species interacts (be they resources, mutualists, competitors, predators, pathogens or parasites), continually modifies the environment in which it lives, exerting selective pressure for evolutionary responses. Because many characteristics of the ecology and life history of a species have consequences for the area and extent of its geographic range, changes in these are liable to affect range size.

If the dynamics of species' geographic range sizes are primarily driven by climatic changes or in response to the demands of the Red Queen, it seems

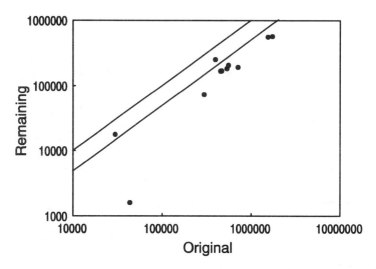

Figure 5.5 Relationship between the sizes of the remaining and the original geographic ranges (km^2) of 11 primate species in South-East Asia. The upper diagonal line represents equality of the two, the lower is where the remaining range is 50% of the size of the original. (From data in MacKinnon and MacKinnon, 1986a.)

unlikely that any simple general model of change in range size through time can be derived. Nonetheless, such models have been proposed, and continue to be debated. Of these perhaps the most enduring has already been mentioned in Chapter 2, Willis' 'age and area' hypothesis (Willis, 1922). This states that the size of a species' geographic range is a function of its age, thus young species have small geographic ranges, while older species have large ones. This model has been roundly dismissed by many authors (see Stebbins and Major, 1965; Fiedler, 1986 for references), largely with reference to examples of species either known to be very old and to have small geographic ranges (paleoendemics), or to be very young and to have large geographic ranges. General inter-specific models can, however, seldom effectively be refuted by simply demonstrating that individual species do not fit, and the age and area hypothesis continues to find support. Thus, McLaughlin (1992) has argued that 'the conclusion that there is a *correlation* between . . . age and range in a sample of species, is inescapable. But to say that the *average* age of narrow species is lower than the *average* age of more widespread species is not the same as saying a species' range is a direct indication of its age'.

The 'age and area' model of the temporal dynamics of geographic range size essentially requires that the bulk of a species' life span (as distinct from the life span of an individual of that species) is spent in range size increase, with the period of decline from maximum range size to extinction being comparatively brief. Long-term stasis in species' geographic ranges is a yet simpler model of range dynamics. It gains support from work on molluscan faunas (Jablonski, 1987 and references therein). This has concluded: (1) that species achieve their full geographic ranges relatively early in their histories (also documented for Foraminifera; Buzas and Culver, 1991); (2) that although some temporal fluctuations in range size occur these are relatively minor; and (3) that the periods of species' origination and extinction encompass the major changes in range size and are both comparatively brief.

If dynamism in species' geographic range sizes was high, there would be no correlation between the range sizes of related extant species, nor between those of ancestral and descendant taxa. Jablonski (1987) has, however, claimed provocatively that geographic range size is a trait that is heritable at the species level. Correlation tests indicate that closely related species of late Cretaceous molluscs tend to have geographic ranges that are more similar in size than expected by chance. The strength of the correlations, it was argued, is comparable with those found between related organisms for many other traits. Combining between-species variation in geographic range size, differential survival of species with different range sizes (see below), and heritability of range size would create the conditions necessary for geographic range to be subjected to species-level selection (Jablonski, 1987). There are, however, some problems. It is, for instance, difficult to determine whether differential survival is directly a consequence of differences in geographic range size or whether species with larger ranges are observed to survive longer because they share other traits which are responsible for their survival. The proportion of variance in range size explained by differences in some traits is low (Jablonski, 1987), but geographic range size is correlated with many traits (Chapter 6).

Further support for stasis in range sizes comes from Ricklefs and Latham's (1992) study of the generic ranges of plants. They find a significant positive correlation between the areas of the geographical ranges of disjunct genera of herbaceous perennial plants endemic to temperate forests of eastern Asia and eastern North America (Figure 5.6). Similar patterns could not be found for other growth forms, and Ricklefs and Latham (1992) suggest that stasis may be associated with ecological specialization. In the face of rapid environmental change, the narrow breadth of conditions suitable for individual specialist taxa will either persist, in which case these will tend to maintain much the same range size, or will not persist, in which case they will become extinct. In contrast, some part of the breadth of conditions which generalist taxa can exploit will always tend to persist – but how large a part may be variable, resulting in substantial changes in range size? This potentially provides a general avenue by which the rather different viewpoints on the long-term dynamics of species' geographic range sizes can be reconciled. More case-studies are necessary before it can be determined if this is so. Such work would be valuable not only for an understanding of the long-term dynamics of rarity, but also in clarifying some of the implications of phylogenetic effects for studies of rarity (Chapter 2).

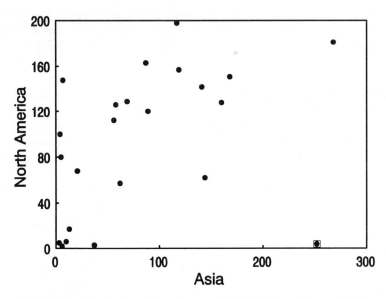

Figure 5.6 Relationship between the areas occupied by disjunct genera of herbaceous plants in eastern Asia and eastern North America. Areas are relative, and estimated from the weight of ranges cut out of photocopied maps. Data from Ricklefs and Latham (1992). The boxed data point was excluded from their analyses as a genus with more than 10 species, in order to reduce the influence of species' number on geographic range.

5.2 RARITY AND PERSISTENCE

The period for which a species persists has been regarded by some as providing an axis, in addition to abundance and range size, along which species can be classified as rare or otherwise, or along which different forms of rarity can be distinguished. The temporal scale considered may be geological (e.g. taxon age; Fiedler and Ahouse, 1992), or ecological (e.g. seasons or decades; Good, 1948). Intuitively, one feels that both abundance and range size are likely to be negatively correlated with persistence; rare species have a lower probability of persisting for relatively long periods. This prediction finds support over both geological and ecological time spans.

5.2.1 Geological time

Analyses of data for marine molluscs and for Foraminifera have revealed that species with large geographic range sizes persist for longer periods of geological time (Figure 5.7; Jackson, 1974; Hansen, 1978, 1980; Jablonski, 1986b, 1987; Buzas and Culver, 1991). This pattern may, however, only hold during periods of normal background extinction. There is evidence that possession of a larger geographic range size did not enhance the survival of species of marine bivalves and gastropods through the mass extinction at the end of the

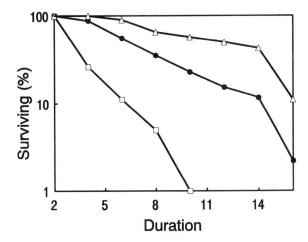

Figure 5.7 The effect of geographic range on the survivorship of Late Cretaceous bivalves and gastropods during periods of background extinction. Symbols represent species of different geographic range size: open triangles >2500 km; closed circles 1000–2500 km; open squares < 1000 km. Duration is expressed in millions of years. (Redrawn from Jablonski, 1986b.)

Cretaceous (Jablonski, 1986b, 1989). The frequency distribution for the geographical range sizes of species within genera surviving the extinction was indistinguishable from that of species within genera which became extinct.

Jablonski (1987) has argued that not only is there a significant relationship between geographic range size and species' duration amongst Late Cretaceous molluscs, but that because species achieved their geographic ranges relatively early in their history, differences in species' durations were, in part at least, a function of geographic range size and not *vice versa*. That is to say, species with larger geographic ranges tended to survive for longer, rather than that species tended to have larger geographic ranges because they had had longer in which to attain them.

These findings have been challenged by Russell and Lindberg (1988a, b), on the grounds that an observed positive relationship between species' duration and geographic range size may simply be an artefact of the data. Taxa having larger ranges have a greater probability of preservation in the fossil record both because they occur at more sites and because they probably tend to have greater local abundances (i.e. local abundance and range size are positively related; Chapter 3). As a result they are likely to appear to persist for longer. In other words the null hypothesis for testing the relationship between species' duration and geographic range size is not a slope of zero. Jablonski (1988) has responded by arguing that if the observed relationship were artefactual it should have a shallower slope for taxa that have a greater potential for preservation as fossils, and that this is not in fact the case for fossil molluscs of the North American Coastal Plain. It remains a moot point, however, how much of the variation in species' durations is actually explained by differences in geographic range size once the effects of sampling artefacts have been removed.

While the bulk of palaeontologists have emphasized the importance of range size in persistence, Stanley (1986) argues that '. . . population size has been of primary importance in determining the vulnerability of species to extinction, whereas geographic range has been of second-order importance.' This assertion is founded upon analyses of Lyellian percentages, i.e. the proportion of species in a fossil fauna which survive to the present, for Neogene marine bivalve molluscs. In virtually all cases the tests are indirect, based on relationships between measures of survivorship and variables which are claimed, though not demonstrated, to be correlated with abundance. The limitations of such an approach, and the criticisms of more direct analyses of the role of geographic range in persistence make it impossible to say whether there are any real differences between taxa in the determinants of their persistence, or whether apparent inconsistencies reflect methodological differences. The question of whether abundance or range size has primacy in determining persistence is, nonetheless, an important one.

5.2.2 Ecological time

Patterns of increasing persistence with greater abundances and range sizes have been documented across ecological time spans (Figure 5.8; Williams, 1964; Terborgh and Winter, 1980; Hanski, 1982a; Diamond, 1984b; Digby and Kempton, 1987; Pimm *et al.*, 1988; Laurance, 1991; Tracy and George, 1992). In the main, studies at these temporal scales have tended to be concerned with local rather than global extinctions. Extinction probability is greater at low population sizes as a consequence of one or more of the following: the Allee effect; stochastic demography (demographic stochasticity and environmental stochasticity); population fragmentation; inbreeding depression; or reduced genetic variation within populations. Demographic factors are probably more important than genetic factors at all but the very smallest population sizes.

As with geological data, such patterns are apt to be strongly influenced by sampling artefacts arising from the greater probability of missing individuals of rare species when they are present than of missing common species. Sampling strategies can be planned so as to reduce this problem in a way which is not achievable with palaeontological data. Notwithstanding, most analyses of persistence in ecological time are based on data sets which were not collected specifically for this purpose, and for which no such approach was taken.

The strength of the link between abundance or range size and probability of extinction is variable, and instances are known in which no simple relationship exists at all (Karr, 1982a, b). This is because a number of other factors influence extinction probability, a topic explored further in Chapter 7.

Whether abundance or range size has primacy in determining persistence is an equally valid question at local scales and ecological time periods as at geographic scales and geological time periods. It is likewise difficult to disentangle.

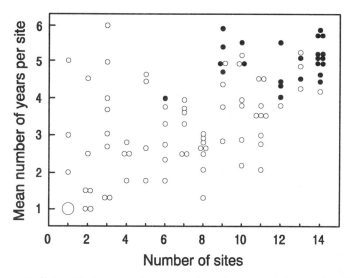

Figure 5.8 Relationship between the temporal persistence and the number of sites at which caught, for moth species sampled at 14 sites over 6 years. Solid circles, average number of individuals >100; open circles, average number of individuals ≤100. (Redrawn from Digby and Kempton, 1987.)

There has been some debate as to whether the generally greater vulnerability of rare species to local extinction because of their low abundance is exacerbated by a greater temporal variability in their local population sizes. All else being equal, it seems reasonable to predict that temporally more variable populations will be at greater risk of extinction because they are more likely to reach very low numbers and be wiped out by chance events (Leigh, 1981; Diamond, 1984b; Pimm *et al.*, 1988). For rare species to be disproportionately disadvantaged by the fluctuations in their populations it would be necessary for them to demonstrate more variable populations on the basis of proportional rather than absolute variabilities. Although there have been many claims that this is the case (see references in McArdle and Gaston, 1992), inter-specific comparisons of measures of variability are severely hampered by problems associated with: equalizing spatial scales; the mean dependence of variabilities; the use of density indices rather than true densities; and the biased nature of some variability measures (McArdle *et al.*, 1990; McArdle and Gaston, 1992; Gaston and McArdle, 1993; Xia and Boonstra, 1992). While the precautionary principle favours the assumption that species with lower abundances have more variable populations this remains empirically insecure. Difficulties in accounting for differences in species' abundances, render direct tests of the link between variability and probability of extinction (Karr, 1982a) essentially uninterpretable.

Measuring real population variability is complicated for rare species by levels of sampling variability. Marquiss (1989) explored the sample sizes

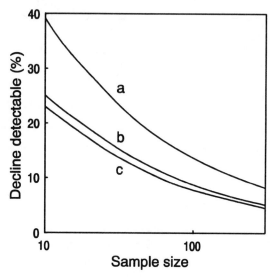

Figure 5.9 The sample sizes necessary to detect population declines of different magnitudes, using three types of heronry census data: (a) counts done in national heronry census years; (b) annual samples in consecutive years involving visits throughout the season and nest inspection; and (c) annual samples when the same observer is usually involved for specific colonies in consecutive years. (Redrawn from Marquiss, 1989.)

required to be able to detect population declines of different magnitudes in the overall population of grey herons (*Ardea cinerea*) breeding in Scotland. Although the species is not particularly rare, this study demonstrates two of the problems in monitoring the populations of species which are rare. Not only was there considerable variation in the sample sizes needed with each of the census methods to detect the same degree of decline in population size (Figure 5.9), but it seemed unlikely that enough heronries would be censused by these methods to detect a decline of less than 10%.

5.3   TRAJECTORIES TO EXTINCTION

Regardless of whether or not there is some general description of the temporal dynamics in the size of a species' geographic range, of its global population or local populations, many rare species are plainly on a trajectory to extinction. Schonewald-Cox and Buechner (1991) proposed simple models encompassing four possible forms this trajectory might take, with respect to changes in the size of a species' geographic range and global population (Wilcove and Terborgh, 1984 explore a related set of scenarios).

(a) Range size remains approximately constant, but the number of individuals declines (Figure 5.10a).
(b) Number of individuals and range size decline simultaneously, such that the density of individuals remains roughly constant (Figure 5.10b).

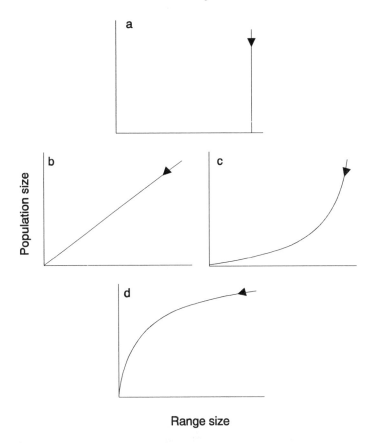

Figure 5.10 Models of the trajectories of abundance and range size towards extinction. See text for details. (Substantially modified from Schonewald-Cox and Buechner, 1991.)

(c) Number of individuals and range size decline simultaneously but the number of individuals declines at a faster rate than range size, such that the density of individuals declines with time (Figure 5.10c).

(d) Number of individuals and range size decline simultaneously but the number of individuals declines at a slower rate than range size, such that the density of individuals increases with time (Figure 5.10d).

We have very little data against which to test these various models, however, it is possible to say something about their relative likelihoods. Model (a) seems improbable in the requirement that collapse in total population size does not affect range size. At the limit, of course, range size cannot remain unaffected.

Model (b) may be relatively frequent where species' ranges are being eroded by habitat loss, but without degradation of the quality of the remaining habitat and without differential loss of high or low quality habitat. The draining of individual ponds, for example, may destroy local populations of aquatic species, thereby reducing their area of occupancy, but may not necessarily have any impact on other local populations.

Model (c) could reflect a situation in which species' geographic ranges are being eroded by habitat loss associated with the degradation of the remaining habitat. The latter may be subtle and difficult to detect. However, fragmentation of habitat could have this effect, with, for example, the increased edge:area ratio of the fragments altering the microclimatic conditions of a far greater area than actually suffers primary habitat loss. It seems probable that this form of trajectory is that which the largest proportion of species are presently on. More generally, it may always have been important for species driven extinct by climate change.

A collapse of range size and population size of the form of model (d) will result when those areas of habitat in which a species fails to achieve higher densities are differentially lost. In principle, successful conservation of the density 'hot spots' of a species' geographic range in the face of the loss of other areas should result in such a pattern of decline. Mayfield's (1983) description of the dramatic decline of Kirtland's warbler (*Dendroica kirtlandii*) suggests that it may represent an example of such dynamics. The range of this species collapsed into its heart, where birds nested as densely as ever.

The trajectories by which species become extinct may be of more than academic interest. Depending on the predominant form of trajectory, different methods of monitoring become most appropriate. Many local populations may, for example, become extinct before major changes are noticed in common indices of geographic range such as the numbers of 10 km grid squares occupied. In this instance the dynamics of range sizes may give a false impression of changes in the vulnerability of a species to extinction (see Hodgson, 1991).

We also need to be aware of the rates at which declines take place and whether these are in any way related to the form of the decline. It is already patently clear that, at least under human influences, dramatic changes can be wrought very rapidly (Figures 5.11 and 5.12). The most rapid decline found by Green and Hirons (1991) was for the Guam flycatcher (*Myiagra freycineti*). In 1981 its world population was estimated at 450 individuals; it was known to be extant in 1983, but was extinct by 1984.

The possible spatial pattern of local extinction of a species is intimately related to the abundance structure of its geographic range (Chapter 4).

5.4   ECOLOGICAL SUCCESS

The overall period of geological time for which a phyletic line survives, relative to other lines that originated at about the same time, can be regarded as the ultimate measure of its ecological success. Wilson (1987) suggests four general causes of phyletic persistence. They are: (a) the number of species in the monophyletic group; (b) the occupation of unusual adaptive zones; (c) the magnitude and numerical fluctuations over time of the effective population size; and (d) the width of the geographic range. Although these criteria cannot be applied with equal ease to individual species, broadly they imply that rare

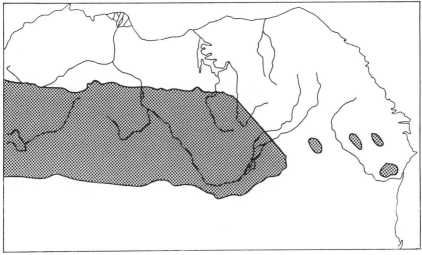

(a)

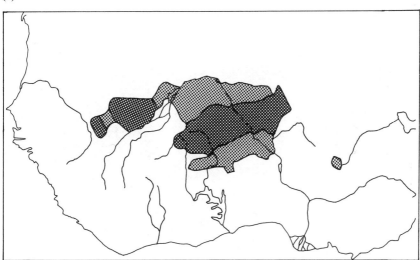

(b)

Figure 5.11 The decline in the geographic range of the korrigum (*Damaliscus lunatus*) in West Africa. A hundred years ago this species probably had the greatest biomass of any West African antelope, and only two species may have been more abundant. Its decline was probably through competition with domestic cattle and human settlement, facilitated by disease. (a) Occurrence in 1900–1920; (b) 1950–1960 (dark hatching indicates areas in which the species probably still occurred but for which no records exist); (c) 1976–1979; and (d) potential range based on the present distribution of its habitat. (Redrawn from Sayer, 1982.)

species will tend to be ecologically comparatively unsuccessful. This line of argument needs to be distinguished from that which says that rare species are

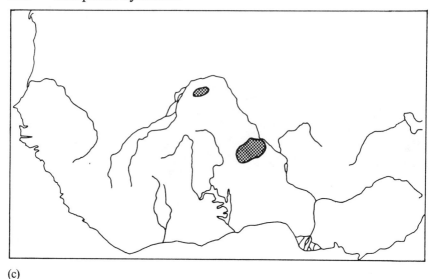

(c)

(d)

Figure 5.11 continued.

failing species, which is a question of the proportion of species which are rare because they are in the declining phase of their range dynamics. One needs also to bear in mind that some species that are presently rare by virtue of their small ranges will, by the criterion of persistence, be regarded as very successful (e.g. tuatara (*Sphenodon*). An important objective is to understand whether such species persisted because they once had much larger geographic ranges, have simply survived by chance, or whether they have evolved adaptations which have resulted in their persistence for periods beyond those typical of species of their range size.

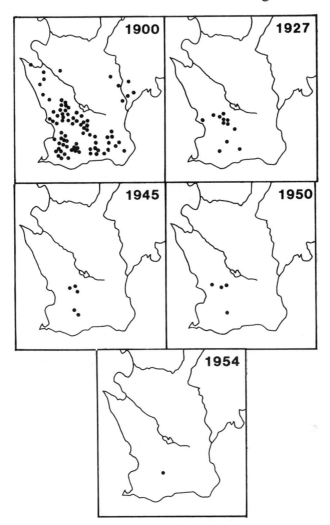

Figure 5.12 The decrease and extinction of the white stork (*Ciconia ciconia*) in Sweden. Dots indicate breeding pairs. The two nestlings of the only 1954 nest died half-grown. (Redrawn from Udvardy, 1969; after Curry-Lindahl.)

5.5   CONCLUDING REMARKS

At the close of the previous chapter regret was expressed that an understanding of the spatial dynamics of rarity was more impressionist than still-life. Knowledge of the temporal dynamics of rarity is on something of a firmer footing. To pursue the analogy further, an understanding of temporal dynamics necessitates moving pictures, and we presently possess substantial fragments of the entire footage. Recreation of the full course of events seems to demand a more insightful splicing of these fragments rather than possession of a larger number of them.

6 Causes of rarity

> Who can explain why one species ranges widely and is very numerous, and why another allied species has a narrow range and is rare?
>
> C. Darwin (1859)

> The causes of rarity are to be found by identifying the constraints on the potential rate at which the population size of the selected species can increase.
>
> J. Greig-Smith and G.R. Sagar (1981)

> It is time to sort out the recent potential causes of rarity, such as fire suppression, direct habitat alteration, increased herbivore populations, or horticultural fancy from the evolutionary consequences of vicariance, genetic depletion, taxon age, aberrant chromosomal events, or possible evolutionary consequences of human intervention.
>
> P. L. Fiedler (1986)

> The list of examples where rare species could lead to erroneous conclusions is limited only by the imagination of the reader.
>
> M. A. Buzas *et al.* (1982)

Several possible causes of rarity have been touched on in preceding chapters (historical factors in Chapters 2 and 5 and environmental factors in Chapter 3). In this chapter, both these and other possibilities are considered in some detail.

At the outset, we need to be clear that there is no reason to believe that it is necessary to invoke a mechanism that is unique to rare species to explain rarity. Rare species 'merely' have lower abundances and/or smaller range sizes than do other species. Both quantities are thus likely simply to be constrained more severely by the same processes which limit the abundances and range sizes of other species. Rarity is not an adaptive strategy, even though an individual organism may benefit from the rarity of the species as a whole.

What is recognized to be a cause of rarity may depend crucially upon the spatial scale and the assemblage with reference to which species have been defined as rare or otherwise. For the purposes of this chapter, 'rarity' will be used somewhat more loosely than has been the case thus far in this book. We will primarily be interested in the factors which cause species to have low abundances and small ranges at meso- and macro-scales when contrasted with reasonably closely related taxa. I suspect that the causes of rarity recognized in individual cases will be more robust in extrapolation to smaller spatial scales than to comparisons involving a broader range of taxa. Regardless, this chapter should be viewed simply as providing a broad framework for understanding the causes of rarity, with extrapolation to fresh comparisons being performed very cautiously.

The chapter comprises six main sections. The first of these concerns some of the approaches by which an understanding can be developed of the causes of rarity. The second and third, respectively, address the two major groups of causative factors, environmental factors (here defined very broadly) and colonization abilities. While not of itself a determinant of low abundances or small range sizes, because it is correlated with many factors that are, body size is discussed separately in the subsequent section. The penultimate and the concluding sections discuss historical considerations and sketch an overview of the causes of rarity. The more artefactual reasons for species being perceived as being rare have already been touched upon (e.g. vagrancy, poor sampling; Chapter 2) and will largely be ignored in this chapter.

6.1   TACKLING CAUSATION

The question of what causes rarity can be addressed in three related ways. The first is to study in detail the biology of individual species which have been categorized as rare and to determine what limits their abundances and range sizes. By concentrating on a single species it may often be possible to perform experiments, many of which need only be very simple, to begin to elucidate mechanisms (Figure 6.1). Such studies of the determinants of rarity of particular species have often been performed, frequently for the purposes of conserving populations (e.g. Roberts and Oosting, 1958; Dring and Frost, 1971; Macior, 1978, 1980; Baskin and Baskin, 1979; Graber, 1980; Moll and Gubb, 1981; Cappucino and Kareiva, 1985; Hutchings 1987a, b; Prober and Austin, 1990; Lesica, 1992). For example, Lesica (1992) studied populations of the aquatic annual plant *Howellia aquatilis* 'to determine life history traits and ecological attributes in order to influence planning decisions on lands managed for multiple use' (the traits and attributes included germination requirements, seed-bank dynamics, effects of substrate on growth, and environmental factors correlated with abundance). Greater emphasis might, in general, usefully be placed on longer term studies.

In attempting a broad understanding of rarity, single species studies provide limited insights. They do not reveal how the factors that determine the abundances and range sizes of rare species differ from those which determine the abundances and range sizes of more common species. For this purpose, comparisons of one or a few rare species with one or a few common congeners (or other close relatives) are most desirable. Sadly, there are very few studies of this kind, especially of animals (examples for plants include Primack, 1980; Pigott, 1981; Fiedler, 1987; Prober, 1992; Karron, 1987). Because they are concerned with several species they are also, unsurprisingly, often somewhat restricted in their consideration of the possible features which may differ between those that are rare and those that are common. Moreover, the increased number of species inevitably reduces the scope for performing manipulative experiments, although the opportunities that do exist to carry them out have not been exploited as fully as they might. The wide ranging

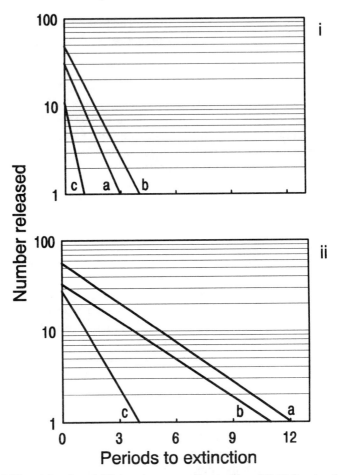

Figure 6.1 Time taken for adult laboratory-reared butterflies of (i) *Heliconius charitonia* and (ii) *H. melpomene* released on a study site to become extinct there, for pulse experiments carried out in three separate years: (a) 1984, (b) 1985 and (c) 1986. Periods are 7-day intervals. (Redrawn from Gilbert, 1991.)

nature of the comparative studies of rare and common North American prairie grasses performed by Rabinowitz and her colleagues is a notable example of what can be achieved (Rabinowitz, 1978, 1981b; Rabinowitz *et al.*, 1979, 1984, 1989; Rabinowitz and Rapp, 1981, 1985; Landa and Rabinowitz, 1983).

Until it is possible to collate the results of a large number of them, detailed comparative studies, each of a few species, will not enable an assessment to be made of the relative frequency with which different factors play a role in causing rarity. To explore this topic necessitates a third, and much cruder, approach. That is the search for broad relationships between species' abundances or range sizes and factors postulated to cause rarity. Such correlative studies can be based on reasonable numbers of species, but are limited to consideration of a few factors chiefly because the data used are usually extracted from the literature rather than directly gathered for the purpose (but see Hodgson, 1986a, b, c, d for perhaps the best illustration of what can be achieved). Experimental support for any conclusions can only realistically be based on a smaller subset of the species (i.e. the previous approach). Broad patterns may anyway often be sought at spatial and temporal scales at which experiments could not be performed.

Common to all three approaches to studying the determinants of rarity are the problems of discriminating between patterns which reflect causal relationships and those which do not, and between causes and effects. Although they have scarcely been considered in this context, phylogenetic constraints may potentially markedly influence observed patterns.

6.2   ENVIRONMENTAL FACTORS

In the absence of any other factors, the abundances and spatial distributions of all species would be limited by environmental variables (abiotic and biotic factors are not discriminated between here, because for practical purposes they often cannot be separated). It thus comes as little surprise to find that there are numerous examples of species which are rare because one or more environmental factors constrain their abundances to low levels or restrict their ranges to small sizes (e.g. Mason, 1946a, b; Stebbins, 1978a, b; Kruckeberg and Rabinowitz, 1985; Woodward, 1987; Prober and Austin, 1990). A review of these case studies would be little different from a general review of the determinants of abundances and range limits, of which a number already exist (e.g. Andrewartha and Birch, 1954; Krebs, 1978; Begon *et al.*, 1990).

One can, of course, ask whether there is any tendency for the abundances and range sizes of rare species to be predominantly constrained by particular factors or combinations of factors. For example, in a study of the threatened vascular plants of Finland, Lahti *et al.* (1991) produced an explicitly subjective evaluation of the probable factors causing the rarity of each species (Table 6.1). This emphasized the importance of climatic and edaphic factors. The value of such analyses is, however, severely limited on three counts. First, making them anything more than subjective is exceedingly difficult. Demonstrating the causal role of particular factors in the rarity of each species would almost certainly necessitate experimental manipulations in each case. Indeed, more generally, studies of the determinants of species' abundances and spatial distributions have provided correlational rather than causal answers. Second, the rarity of many species is unlikely to be a result of just one factor. The rarity of the pierid butterfly (*Pieris virginiensis*) for example, has been shown to be a product of unpredictable host phenology, unfavourable conditions during the oviposition period, poor abilities to lay eggs rapidly, poor dispersal and the effects of granulosis virus (Cappucino and Kareiva, 1985). Third, it seems unlikely that the relative roles of different environmental factors in causing

Table 6.1 Summary of the factors presumed to affect the rarity of threatened vascular plant species in Finland. (From Lahti *et al.*, 1991.)

	Category						
	Extinct		*Endangered*		*Vulnerable*		*All*
Number of taxa (%)	7 (100)		33 (100)		43 (100)		83 (100)
Climate	7 (100)		25 (76)		32 (74)		64 (77)
Edaphic factors	1 (14)		17 (52)		27 (63)		45 (54)
Dispersal	3 (43)		2 (6)		1 (2)		6 (7)
Establishment	– (–)		6 (19)		3 (7)		9 (11)
Population longevity	1 (14)		1 (3)		– (–)		2 (2)
Hybridization	– (–)		3 (9)		3 (7)		6 (7)

the rarity of species in a given assemblage could be generalized to other assemblages in anything but the vaguest fashion.

Rather than dwell on the various environmental factors which are important in causing the rarity of individual species, we can ask a more fundamental question. Are rare species rare because they are only capable of exploiting a narrow range of environmental conditions, or because the spatial extent of the conditions they can exploit is highly restricted, or both? A formal framework in which this question can be addressed is provided by the concept of niche pattern (Shugart and Patten, 1972). This is a means of summarizing a great deal of ecological information about a species' assemblage, and involves plotting in three-dimensional space the coordinates of species' niche breadths, niche positions and abundances. Niche breadth is the range of environmental conditions (which can be defined broadly to include habitat factors and food types) in which a species is observed to survive and reproduce. Niche position has been interpreted in various ways, for example as a measure of the actual availability of these conditions, or of how typical they are of the universe of conditions under consideration (Shugart and Patten, 1972; Mac Nally and Doolan, 1986; Seagle and McCracken, 1986). Counter-intuitively, the greater the value of niche position the less available or the more atypical the conditions are.

Perhaps inevitably, agreement that developing an understanding of niche pattern is important has arisen far more rapidly than has the understanding itself. There have been various attempts to examine the interactions between one or more pairs of the three variables (e.g. Shugart and Patten, 1972; Dueser and Shugart, 1979; Mac Nally and Doolan, 1986; Seagle and McCracken, 1986; Burgman, 1989; Mac Nally, 1989; Shenbrot *et al.*, 1991; see also McNaughton and Wolf, 1970; Ricklefs, 1972; Hanski and Koskela, 1978; Adams and Anderson, 1982). Unfortunately, methodological difficulties have rendered some of this work difficult to interpret. The following conclusions are therefore based solely on some of the more recent analyses which have attempted to deal with these problems. These find that: (a) abundance and niche breadth are not significantly correlated (Seagle and McCracken, 1986; Robey et al., 1987; Burgman, 1989; Mac Nally, 1989; Shenbrot et al., 1991); (b) abundance and niche position show either significant negative correlations or no relationship at all (Seagle and McCracken, 1986; Robey et al., 1987; Mac Nally, 1989; Urban and Smith, 1989); and (c) niche position and niche breadth sometimes show negative relationships and sometimes positive relationships (Mac Nally and Doolan, 1986; Robey et al., 1987; Mac Nally, 1989). In short, there is little evidence that rare species have niches that are narrower than those of abundant species, but some evidence that the sets of conditions which fall within the niche space are more scarce or unusual in physical space for rare species than they are for abundant ones.

The concept of niche pattern was developed in the context of niche measures based on a set of variables reduced from a larger set of continuous multivariate data. Other considerations of niche measures have, however, been based on data which are discrete, or treated as such. They cannot be reduced in the same way, and using them, niche axes may have to be considered independently. Moreover, although methods have been developed to account for differences in the availability of different resource states (e.g. Smith, 1982), some analyses of discrete data continue to ignore such considerations and to rely on the original formulations of niche measures (chiefly those of Levins, 1968). A number of studies have explored relationships between niche breadths along individual axes measured using discrete data, and species abundances or range sizes (e.g. Price, 1971; Parrish and Bazzaz, 1976; Müllenberg et al., 1977; Ford, 1990; Shkedy and Safriel, 1992). The results have been mixed and, for the reasons mentioned, are often difficult to interpret. They seem to provide little basis for arguing that, in general, abundant and widespread species have broader niches.

A number of studies not explicitly concerned with niche measures have documented positive relationships between the abundances of species and the relative availability of the resources they exploit (e.g. Hodgson, 1986a; Dixon and Kindlmann, 1990; Gilbert, 1991).

The conclusion that there is rather limited evidence for a strong relationship between abundance or range size and niche breadth, as against resource availability, appears at odds with an often repeated explanation for the observed positive relationship between the local abundance and meso- or macro-range size of species within an assemblage. This is that species that can exploit a wide range of conditions locally, and in so doing achieve high densities, will also be able to survive at more sites, while those that can only exploit a narrow range of conditions will be unable to attain either high local densities or extensive occurrences (Chapter 3; Brown, 1984). Clearly this explanation rests on the assumption that niche breadth and abundance are positively related. There are several possible reasons why this conflict has arisen (summary list in Table 6.2).

Table 6.2 Factors possibly explaining or contributing to the conflict between the poor evidence for a strong relationship between niche breadth and abundance or range size, and the claim that such a relationship explains why locally abundant species tend to be widespread. See text for details

- Effects of spatial scale
- Sampling artefacts
- Individual-orientated *v.* species-oriented views of niche breadth
- Link between niches and habitats
- Abiotic *v.* biotic niche dimensions
- Vagrants
- Numbers and relatedness of taxa
- Theoretical inadequacies

(a) Spatial scale

Most analyses of niche breadths and positions are carried out over relatively small spatial scales. It may, therefore, simply be that their results do not bear extrapolation to the large scales over which regional and geographic range sizes are measured. There is, however, some evidence to the contrary. The studies of Burgman (1989) and Mac Nally (1989), on plants and birds, respectively, have been performed at large scales. These have failed to find significant relationships between niche breadth and either species' patterns of occurrence (Burgman, 1989) or local abundance (Mac Nally, 1989).

Seagle and McCracken (1986) offer a somewhat different interpretation of the effects of scale on observed relationships between niche breadth and abundance. They point out that while in direct analyses of these relationships, species' abundances and niche breadths are quantified at the same spatial scales, with respect to relationships between abundance and range size, abundance is usually measured locally and range size regionally or geographically. Again this discrepancy may account for the conflict over the causes of the relationship between abundance and range size, although there is no direct evidence that this is so.

(b) Sampling artefacts

Sampling artefacts have been demonstrated to have a profound effect on observed relationships between niche breadth, niche position and abundance. Inevitably, rare species will tend to be recorded from fewer sites than they actually occur at simply because they have been overlooked (a problem that is potentially exacerbated by the short duration of most studies). This will tend consistently to bias measures of niche breadth and niche position (Seagle and McCracken, 1986; Burgman, 1989), and is one of the problems which makes many analyses difficult to interpret. It seems quite plausible that such artefacts may lead to a perception that species which are more widespread and more abundant have wider niches even when this is not actually the case.

Table 6.3 Occurrence of endemics in unique or common habitats, based on data from inventories of rare and endangered plant taxa. (From Kruckeberg and Rabinowitz, 1985.)

	Taxon in unique habitat		Taxon in common habitat		Uncertain		Total (No.)
	No.	%	No.	%	No.	%	
California	560	49	388	34	185	16	1133
Oregon	139	49	111	39	34	12	284
Washington	65	47	74	53	0	0	139
Yukon Territory	5	38	8	62	0	0	13

(c) Individual and species-oriented views

It is important to distinguish between differences in niche breadth that result from the flexibility of individuals and those that result from the requirements of individuals. If a broader niche means that a species can exist in areas that have very different environmental conditions, then range size and niche breadth may well be positively related. If, however, a broad niche means that individuals of a species need areas with a variety of resources to exist, then species with broad niches may actually be more limited in the number of areas they can exploit.

(d) Niches and habitats

It has often been documented that many rare species are restricted to single habitat types (Table 6.3), and that abundances and range sizes tend to increase with the numbers of habitats species occupy (Figure 6.2; Glazier, 1980; Thomas and Mallorie, 1985b; Pomeroy and Ssekabiira, 1990; Pagel *et al.*, 1991; C.D. Thomas, 1991; Kattan, 1992). Both observations have been taken as support for the argument that abundances and range sizes are positively related as a consequence of differences in species' niche breadths. Such evidence needs to be treated cautiously, for three main reasons. First, at large spatial scales, habitat assignations are generally very crude, often being based solely on the overlap between a map of a species' extent of occurrence and a map of the distribution of habitat types. Species' actual habitat associations may be much more specific than the results of such analyses would imply, and there is thus probably a strong artefactual component to them. Morrison *et al.* (1992) provide an excellent review of the measurement of wildlife–habitat relationships. Second, the connection between the numbers of habitat types a species occupies and the size of its niche is not of necessity a simple one. There is no reason to believe that all habitats contribute an equal amount to niche space. Third, relationships between abundance or range size and habitat occupancy are highly susceptible to the sampling artefacts of individual and species-oriented views.

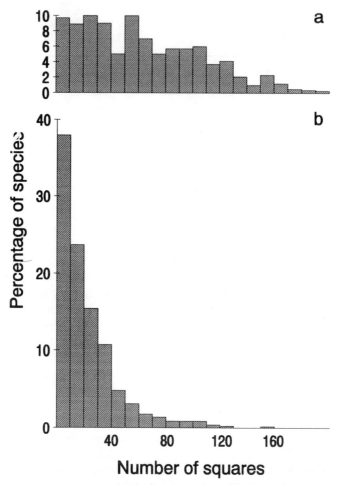

Figure 6.2 Frequency distributions of the geographical range sizes of terrestrial bird species in Africa, divided into (a) generalists and (b) specialists. Specialists have 60% or more of records from one of three possible habitat zones (montane and forest, moist woodland, or dry woodland and desert), while generalists occur more widely. Squares are 2 ½ ° by 2 ½ °. (Redrawn from Pomeroy and Ssekabiira, 1990.)

There are a growing number of examples of independence of range size or population size and habitat specificity (Fowler and Lawton, 1982; Pianka, 1986; Rabinowitz *et al.*, 1986; Hanski *et al.*, 1993).

(e) Abiotic and biotic niche dimensions

Most analyses of differences in species' niche breadths and niche positions have concentrated on habitat variables, and often primarily abiotic ones. The range of states of these variables at which species occur may be markedly influenced by biotic interactions, resulting in an implicit incorporation of biotic variables into analyses. However, it remains unclear how explicit incorporation of biotic variables reflecting the full variety of species–species interactions (e.g. consumer–host, predator–prey, mutualist) might alter perceptions of differences in species' niche statistics. It is noteworthy, nonetheless, that strong predator and parasite pressure undoubtedly accounts for the low abundances and small range sizes of many species (e.g. Monro, 1967; Lubchenco, 1978; Room *et al.*, 1981; Estes *et al.*, 1982; Case and Bolger, 1991; Hegazy and Eesa, 1991), and that the notion that competitive inferiority typifies many rare species has long been held (e.g. Griggs, 1940; Mitchley and Grubb, 1986; but see also Rabinowitz *et al.*, 1984). Clarification of the relationship between the rank order of species' niche breadths determined on the basis of habitat variables, and of a more inclusive set of variables would be valuable.

(f) Vagrants

A concern which too often appears to be entirely ignored in deriving niche statistics is that of the treatment of vagrant individuals (defined broadly, as in Chapter 1). These may have a profound effect on niche statistics, because they may be found living in conditions under which they cannot reproduce, or under which populations are not self-sustaining (sink populations). Strictly speaking, these conditions cannot be regarded as being part of a species' niche, the breadth of which they will tend to inflate. Even very large populations may not be self-sustaining (Pulliam, 1988), raising the possibility that most individuals of some species may be found in conditions which lie strictly outside of the species' niche space. Unfortunately, vagrants are often difficult to recognize as such. We do know, however, that their significance is likely to be a function of spatial scale, with the contribution of vagrancy to species' richness probably peaking at meso-scales (Chapter 1). Thus vagrancy may differentially inflate apparent niche breadth at these scales.

(g) Numbers and relatedness of taxa

Relationships between the numbers of habitats species occupy and their abundance or range size, and between the latter two variables themselves have frequently been based on large numbers of sometimes quite distantly related species. In contrast, analyses of niche pattern have often been restricted to a few, more closely related, taxa. Such differences in the numbers and relatedness of species may have profound effects on the relationships which are observed, although it must be said it is not entirely obvious in precisely what way.

(h) Theoretical inadequacies

Theoretically it seems more reasonable to expect a positive relationship between a species' range size and its niche breadth than between its abundance and niche breadth (Kouki and Häyrinen, 1991; Hanski *et al.*, 1993). If local abundances and niche breadths were correlated, there would seem little advantage to specialization; the jack-of-all-trades is master of all. On the other hand, one might expect to find a relationship between niche breadth and range size, because this is not subject to individual selection.

It is premature to conclude whether rarity is predominantly caused by the availability of resources or the breadth of resources which can be utilized. Indeed, there need be no simple answer. However, there is certainly more support for low abundances and small range sizes resulting from a lack of resources and less for them being generated by differences in the breadth of resource usage than is commonly supposed. Moreover, there are ample reasons to doubt the generality of a niche-breadth based explanation for abundance–range size relationships.

6.3 COLONIZATION ABILITY

If the physical distribution of the conditions delimited by a species' realized niche and its pattern of spatial occurrence are entirely congruent, then abiotic variables and biotic interactions would seem sufficient to explain its range size. However, if there are places which could be mapped into the species' niche space, but at which it does not occur, then clearly we will need to invoke some further constraints. Such constraints can be drawn together under the broad heading of colonization ability.

In fact there have been rather few direct attempts to ascertain the extent of this matching. There is, however, growing interest in determining species' 'bioclimatic envelopes' (the sets of combinations of climatic conditions under which they are found; Busby, 1986; Longmore, 1986; Murray and Nix, 1987; Walker, 1990; Lindenmayer *et al.*, 1991; J.E. Williams, 1991). This has been engendered by the need to predict species' present distributions in areas where sound distributional data are lacking, and to predict how these distributions are likely to respond to climate change. Such data provide an excellent opportunity to examine the relationship between the geographical distribution of species and the geographical distribution of the conditions which lie within their niche space.

It is evident that the spatial distributions of many species are limited by their colonization abilities. The success of intentional and accidental introductions of species into areas where previously they did not occur is sufficient demonstration. There is therefore little to be gained by asking whether there are rare species the ranges of which are limited by colonization abilities. The answer is an emphatic yes. A more interesting question is whether rare species tend on average to have poorer colonization abilities than common species. This can be addressed with respect to each of the two major components of colonization ability, that is dispersal ability (how well a species can reach new sites) and establishment ability (how well it can establish itself once there).

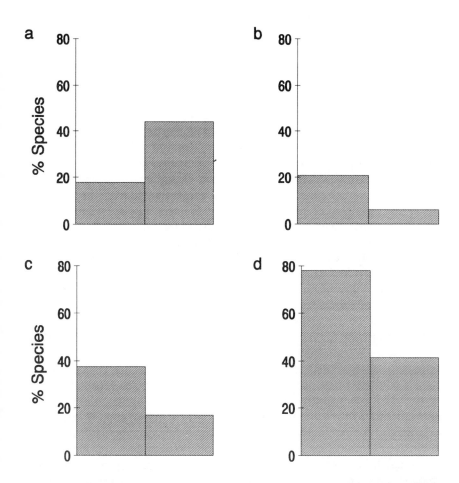

Figure 6.3 The percentage of common British plant species with different dispersal strategies which are increasing or decreasing in contemporary landscapes. In each instance the left hand column is the percentage of species increasing, while the right hand column is the percentage decreasing. (a) No strategy for dispersal in time or space detected; (b) numerous widely dispersed seeds; (c) dispersal of shoot or root fragments important; and (d) persistent seed bank. (Redrawn from Hodgson and Grime, 1990.)

6.3.1 Dispersal ability

The ranges of a variety of species have been found to be limited by dispersal (e.g. Carter and Prince, 1988; Lawton and Woodroffe, 1991; Primack and Miao, 1992). Other studies have attempted to examine how dispersal ability and range size interact. However, because it is difficult to measure, they have in the main either ignored the fact that dispersal ability is a continuous variable, and have contrasted sets of species thought to differ markedly in dispersal ability (e.g. winged *v*. brachypterous, planktonic *v*. nonplanktonic), or they have sought supposed correlates of dispersal ability (e.g. size of dispersal stage). Neither approach is entirely satisfactory.

On the basis of the published literature, it would seem that there is a positive relationship between a species' dispersal ability and range size within a taxonomically defined assemblage (e.g. Hansen, 1980; Reaka, 1980; Juliano, 1983; Kavanaugh, 1985; Söderström, 1989; Hedderson, 1992; Oakwood *et al.*, 1993). Kavanaugh (1985), for example, reported that for Nearctic carabid beetles of the genus *Nebria* the mean range size of brachypterous species was only 14% of that for macropterous species. The dispersal ability–range size relationship may often be asymmetric, with good dispersers having small, intermediate and large range sizes, but poor dispersers tending only to have small to intermediate ranges.

Human activities have a profound effect on species spatial distributions, and may complicate attempts to ascertain whether simple relationships exist between rarity and other ecological parameters, and if so to explain them. This is particularly evident when seeking possible relationships between range size and dispersal ability. Human-modified or maintained habitats now dominate large regions of the earth, and it is not surprising therefore that common species tend to be associated with them. These same habitats are frequently temporally unstable (e.g. agricultural lands), and common species tend also to be more frequently associated with unstable or disturbed habitats than do rare species (e.g. Glazier, 1980; Hodgson, 1986a; C.D. Thomas, 1991). Occupation of such habitats often necessitates possession of well-developed dispersal abilities, effectively selecting for a relationship between range size and dispersal ability. Such a relationship may not have existed were the relative extents of disturbed and comparatively undisturbed habitats different, and hence dispersal ability may not itself be a cause of rarity. Equally, abundant and widespread species may have possessed better dispersal abilities and been pre-adapted to exploit increases in the relative extent of disturbed lands. Many species that are presently rare are members of climax assemblages whose areal extent has been severely reduced in recent times. However, their previous extent may not have necessitated the high dispersal abilities which presently seem desirable to ensure recolonization of patches where the species become extinct. Whatever, species with effective dispersal abilities currently have, in general, increasing range sizes, while those with poor dispersal abilities are decreasing (Figure 6.3; Turin and den Boer, 1988).

Even where we accept that rare species tend to have poorer dispersal abilities than common species, we must be wary of relying on a single explanation. While it seems probable that in some instances poor dispersal abilities cause rarity, in others they may be a response to rarity. There is little advantage to be had from well-developed dispersal abilities if, for example, the probability of encountering unoccupied yet suitable sites is very low. Indeed, selection for a reduction in dispersal abilities may in some instances be very rapid when opportunities for encountering suitable sites decline (Dempster, 1991).

Table 6.4 Relationships reported by Glazier (1980) to exist between the range size and other traits of North American *Peromyscus* mice (many were shown to hold within subgenera and species' groups as well as across the genus as a whole)

Species with small ranges tend to:

Occur	● less frequently in marginal habitats
	● over a narrower altitudinal range
Have	● smaller average litter size
	● smaller maximum litter size
	● greater longevity
	● greater body size
	● smaller fecundities
	● lower reproductive effort (energy cost of raising a single litter)

6.3.2 Establishment ability

In principle the distinction between rarity that is determined by environmental factors, and that as a result of colonization abilities is reasonably clear. In practice the division is substantially more blurred. In particular, it may be very difficult to distinguish between establishment that fails because of a poor match between the environmental requirements of a species and those of a given site, and that which fails because the population of a species cannot grow fast enough to avoid extinction through stochastic processes.

A good demonstration of this point has been provided by the debate over the determinants of the success or failure of introductions of species into sites (and often continents) where they did not previously exist (Drake *et al.*, 1989). For several systems, positive correlations have been found between species' native abundances and/or range sizes, and the success of their introduction (e.g. Forcella and Wood, 1984; Moulton and Pimm, 1986; Forcella *et al.*, 1986; Hanski and Cambefort, 1991; Roy *et al.*, 1991). Success can be scored simply as success versus failure, or in terms of the range sizes and abundances species achieve in the area to which they were introduced. How general such examples are can be disputed, but the point remains that, assuming they are not artefactual, it is exceedingly difficult to ascertain whether differences in success result from: (a) differences in the matching of the climate and other environmental factors of the native ranges of the species and the area to which they were introduced; (b) differences in the population growth rates of the species; or (c) both.

One may, of course, simply ask whether there is any evidence that rare species have reproductive characteristics which, ignoring climatic matching, might indicate a poor establishment ability. Rarity and characteristics of reproductive biology have been associated in studies of a variety of taxa (e.g. Meagher *et al.*, 1978; Harper, 1979; Glazier, 1980; Hodgson, 1986b; Spitzer and Lepš, 1988; Karron, 1987; Wiens *et al.*, 1989; Paine, 1990; Longton, 1992;

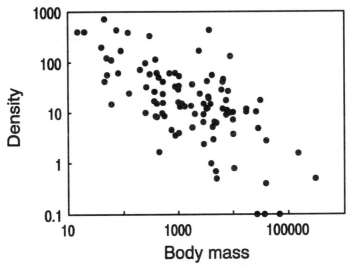

Figure 6.4 Relationship between the density (individuals/km^2) and the body mass (g) of Neotropical forest mammals. (From data in Arita *et al.*, 1990.)

Sutherland and Baillie, 1993). Glazier (1980), for example, found correlations between range size and a variety of traits of North American *Peromyscus* mice species (Table 6.4). In this instance, features, such as greater litter size and greater fecundity, which one might expect to increase probability of establishment were associated with species with larger geographic range sizes. Likewise, in the Sheffield flora a smaller proportion of rare species than common species have readily germinating seeds, or a capacity for vegetative lateral spread, while a higher proportion produce no viable seed in the study region (Hodgson, 1986b).

Most explanations of observed relationships between rarity and reproductive traits invoke some environmental constraint, such as intense competition, as forcing rare species to sacrifice traits associated with a high establishment ability in order to survive. If these interpretations are correct, then the effect of a low probability of establishment on rarity would be a secondary cause stemming from the environmental constraints.

6.4 BODY SIZE

In considering the causes of rarity, a theme that has thus far been ignored is the role of differences in the body sizes of species. Although body size cannot be regarded as a proximate cause in the same way as environmental factors and colonization abilities can, it is related to so many ecological characteristics of species (Peters, 1983) that it deserves some attention in its own right.

The interaction between rarity and body size is itself a complex one. For a number of data sets, abundance and body size are negatively correlated; large

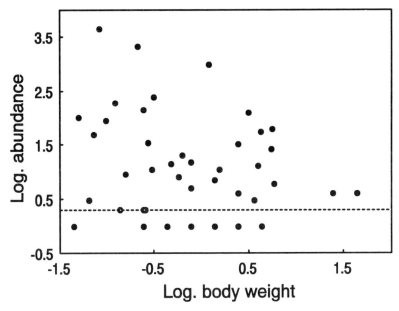

Figure 6.5 Relationship between the abundance and body weight (mg) of the inverte-brate fauna of a freshwater pond in northern England. Dashed line delineates those species which under a quartile definition are categorized as rare. (Redrawn from Blackburn *et al.*, 1993.)

species tend to be rare (Figure 6.4; e.g. Damuth, 1981, 1987; Peters, 1983; Peters and Wassenberg, 1983; Peters and Raelson, 1984; Marquet *et al.*, 1990; Currie, 1993). For many others, there is either no simple interaction between abundance and body size or an approximately triangular one; rare species do not fall in a particular size class (Figure 6.5; e.g. Brown and Maurer, 1987; Morse *et al.*, 1988; Maurer *et al.*, 1991; Blackburn *et al.*, 1993). Finally, within taxa of low rank (e.g. genera and tribes as opposed to families or orders), abundance and body size are sometimes positively related; rare species are small (Nee *et al.*, 1991c). By and large, it would seem that these different results are not at odds with one another, but can be reconciled with reference to such factors as the between-species range of body sizes, whether abundances reflect mean or maximum and crude or ecological densities, and the taxonomic extent of an assemblage (Lawton, 1989; Carrascal and Tellería, 1991; Nee *et al.*, 1991c).

The relationship between range size and body size seems to be rather simpler than that between abundance and body size, however, one suspects that this may be because, in contrast, it has received comparatively little consideration. In general, range size is an increasing function of body size, or at least the minimum range size of species in a body size class increases with the mean body size of that class (Figures 6.6 and 6.7; Van Valen, 1973b; Reaka, 1980; Brown, 1981; McAllister *et al.*, 1986; Brown and Maurer, 1987, 1989; Arita *et*

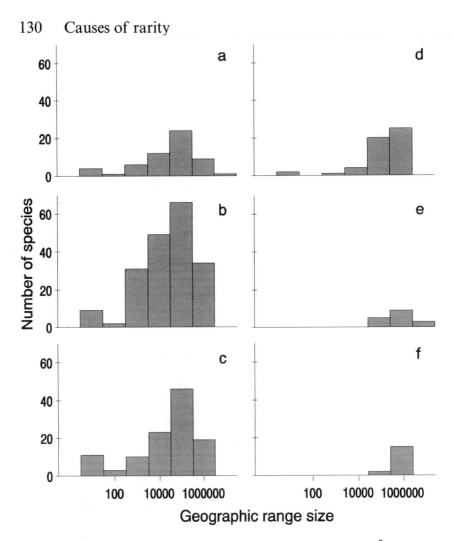

Figure 6.6 Frequency distributions of the geographic ranges sizes (km^2) of North American land mammal species, divided into body size classes. (a) <16 g, (b) 16–128 g, (c) 128–1024 g, (d) 1024–8192 g, (e) 8192–65 536 g and (f) > 65 536 g. (Redrawn from Brown and Nicoletto, 1991.)

al., 1990; Brown and Nicoletto, 1991; Maurer *et al.*, 1991; Ayres and Clutton-Brock, 1992). Put another way, species tend to fall within a roughly triangular region of range size–body size space, with species with small range sizes tending to have small body sizes. Although some studies have failed to find significant correlations between range size and body size (e.g. Juliano, 1983), these are in the main difficult to interpret as results may conform to the above pattern without yielding such correlations. Figure 6.8 suggests, nonetheless, that there is little evidence of any marked pattern in some data. Glazier (1980) reports a negative interaction between range size and body size for North

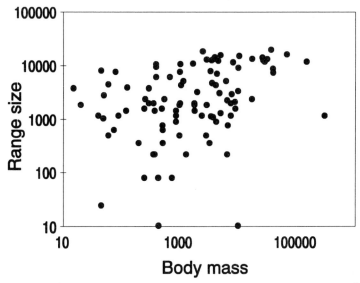

Figure 6.7 Relationship between the geographic range size (thousands of km²) and the body mass (g) of Neotropical forest mammals. (From data in Arita *et al.*, 1990.)

American *Peromyscus*, and positive interactions, negative interactions and the absence of relationships within individual species' groups of the genus. These results suggest the likelihood that there is a taxonomic effect in range size–body size relationships akin to that for abundance–body size relationships.

Work on both abundance–body size and range size–body size patterns and proposed mechanisms has concentrated upon animal taxa. Doubtless, in major part, this is because of difficulties in attributing relative body sizes to plant species (because of their plastic growth). There are, nonetheless, some examples of similar patterns for plants. Aizen and Patterson (1990) find a positive relationship between the range size and height of oak trees. Harper (1979) favours the view that larger (woody) plant species are in general under-represented in the rare component of floras, while Oakwood *et al.* (1993) find that plant species of taller growth forms tend to occupy a smaller number of regions of Australia (Figure 6.9). More work of this kind may be important in understanding mechanisms.

For animals, explanations of the interactions between species' abundances or range sizes and their body sizes, have centred principally on energetic constraints (Damuth, 1987; Brown and Maurer, 1987, 1989; Maurer and Brown, 1988; Root, 1991). Because the per capita resource requirements of species are a function of their body sizes, trade-offs have been invoked between species' body sizes and the densities they can attain. As a result of these trade-offs, it has been argued, larger species will only have total population sizes that are viable in the long term if they have larger geographic ranges. It

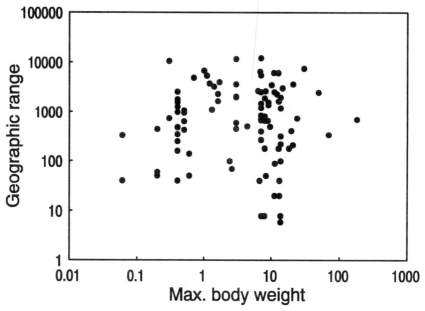

Figure 6.8 Relationship between the size of the geographic range ($\times$ 1000 km^2) and the maximum body weight (kg) of species of primate. (From data in Wolfheim, 1983.)

has steadily become more apparent that such lines of reasoning are, however, inadequate (Lawton, 1989; Nee *et al.*, 1991c). They undoubtedly do not explain some of the abundance–body size patterns that are observed, and it remains a topic of debate whether they explain any at all. Increasingly it has seemed that there is little prospect of understanding abundance–body size and range size–body size relationships without first developing a thorough under- standing of the determinants of the forms of the frequency distributions of species' abundances, range sizes and body sizes (Blackburn *et al.*, 1990; Gaston *et al.*, 1993). In the context of this chapter, this brings us full circle, back to the causes of rarity.

It may not be straightforward to dismantle the problem in this way. Many of the factors that have been mentioned as playing a role, if not always alone, in causing rarity have been claimed to be related in some way to body size. These include species environmental tolerances (Wasserman and Mitter, 1978; Cawthorne and Marchant, 1980; Lindstedt and Boyce, 1985), dietary niche breadths (Hansen and Ueckert, 1970; Jarman, 1974; Wasserman and Mitter, 1978), dispersal abilities (Southwood, 1981; Rapoport, 1982), and rates of increase (Fenchel, 1974). In many instances these interactions are not simple. They may, for example, be non-linear (e.g. dispersal abilities; Southwood 1981) or may depend on taxonomic level, perhaps increasing with body size for closely related species but decreasing for more distantly related species (e.g.

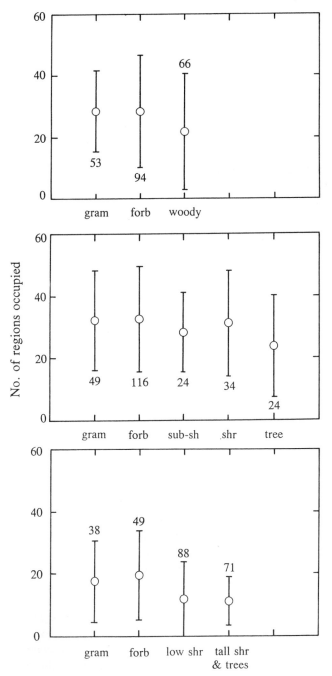

Figure 6.9 Mean (± 1 SD) number of regions (of a total of 97) occupied by native species within different growth forms for local floras from three regions of Australia: (a) central Australia, (b) western New South Wales, and (c) Sydney. Gram – graminiods; sub-sh – subshrubs; shr – shrubs; tall shr – tall shrubs. (Redrawn from Oakwood *et al.*, 1993.)

rates of increase: Fenchel, 1974; Williamson, 1989b; Stemberger and Gilbert, 1985). In considering the role of body size in rarity, establishing causality is, once again, difficult.

6.5 HISTORICAL CONSIDERATIONS

Emphasis has been laid on causes of rarity operating at the time at which this status was determined. Species may, however, be rare because of events which occurred (or did not occur) in the past. These are often not readily deduced, and when determining the proportion of species in an assemblage whose rarity is the product of different processes, historical factors may frequently account for the tally of species for which causes remain obscure.

Niche statistics determined at any one time may not adequately reflect differences between species which might be observed were these measurements to define the niche space within which successful reproduction could occur. For some long-lived species, conditions for reproduction only arise occasionally in periods perhaps of many years.

Species may also be rare in terms of having small ranges because they are limited by their colonization abilities, without their necessarily being poorer colonists than more widespread species. Historical factors, such as their recent speciation or the environmental conditions they experienced in the past, may mean that species have had less time to spread into areas which they have the capacity to occupy at the present. This is the essence of the age and area hypothesis (Chapter 5).

6.6 OVERVIEW

A number of observations can be made, by way of a summary of our present understanding of the causes of rarity:

(i)   The causes of rarity are by and large idiosyncratic, and beyond the broadest of generalizations it is impossible to predict in advance the reasons why any one species is rare. Although rarity has been regarded by some as a syndrome typified by a defined set of traits, there seems little evidence to support such a belief. Concern to relate particular traits with the state of being rare has obscured much of the heterogeneity in the phenomenon (Rabinowitz, 1981a).

(ii)   Distinguishing the causes of rarity from its effects poses a considerable obstacle. This situation is aggravated because not only can the causes of rarity of one species be the effects of rarity for another, but in some instances effects have the potential to become causes. A poor dispersal ability, for example, may be an effect of rarity which results from selection on the individual to avoid investing resources in dispersal mechanisms which are of no value because there are no sites to be colonized. If the potential to evolve a good dispersal ability is not retained, however, changes in the environment which generate many potentially suitable sites may not be exploitable and dispersal ability may become the cause of rarity. The ease with which changes in dispersal ability, in particular, are selected for, may become an important determinant of the effects of global climate change upon extinction rates.

(iii)   The causes of rarity can be investigated at a variety of levels. In the main this chapter has been concerned with broad ecological determinants of rarity. However, a comprehensive understanding of the causes of the phenomenon also requires knowledge of the genetic dimensions to rarity and the mechanisms by which the ecological determinants take effect. The former considerations lie beyond the scope of this book (Fiedler (1986) provides a useful entry into the relevant literature). The latter remain little explored. Thus, while we know that environmental factors restrict species' distributions, we have only a limited idea how in practice these interact with the components of a species' population dynamics to do so. Beddington *et al.* (1976) suggest three possible changes that might occur toward range margins. First, changing environmental conditions might alter average population dynamics such that no stable population dynamics exist. Second, unpredictable environmental events may attain levels such that the probability of a population being pushed to extinction tends to unity within a few generations. Third, the time taken by a population to return to equilibrium levels may become so long that small perturbations in population size may lead to extinction.

(iv)   The causes or postulated causes of rarity are dependent upon both spatial and temporal scales. Carter and Prince (1988) cite the example of the plant *Colophospermum mopane*. At the scale of southern Africa at large, the limit of the occurrence of this species correlates with the mean 5°C July isotherm. Within Botswana alone, occurrence is correlated with rocky hills, drainage channels, coarse grained sands and cultivated areas, from which it is absent. On the larger scale the correlation with soils is obscure, while on the smaller scale the correlation with climate is obscure. Hence, in the field the limit always appears to be associated with features of the terrain and the plants' critical responses to climate are masked by larger responses to local factors.

(v)   The low abundances and small range sizes of many species presently regarded as rare are products of human activity. Thus, the major factor regulating commonness and rarity in the Sheffield flora at the present time is land use (Hodgson, 1986a). Species able to exploit artificial, productive habitats are generally common, while species restricted to less productive habitats tend to be rare (Hodgson 1986a, b).

(vi)   The causes of rarity can be viewed as hierarchical. Fiedler and Ahouse (1992) distinguish four categories of rarity, defined by the possible combinations of wide or narrow range size and short or long temporal persistence (taxon age). For each of these categories they identify a hierarchical ranking of the most probable causes of rarity. These include both ultimate and proximate causes. Whether one agrees with their proposals or not, these hierarchies emphasize the lack of any consistent pattern of causation.

(vii)   Most studies of the causes of rarity have concerned only a few of the possibly relevant factors.

7 Conservation and rarity

Suddenly, as rare things will, it vanished.

R. Browning (1855)

... there is unfortunately no precedent for 5 billion human beings suddenly sharing an enlightened vision of the future.

N.R. Flesness (1992)

There are no hopeless cases, only people without hope and expensive cases.

M.E. Soulé (1987)

Considerations of rarity lead almost ineluctably to the topic of conservation. Indeed, it seems a popular belief that the two issues are inseparable. Previous chapters should have established beyond any doubt that this is not so. There are many questions about rarity which are of interest and yet have little directly to do with conservation. Nonetheless, it would be wrong to ignore this, the most important of the applied dimensions to the study of rarity.

The principal justification for a strong link between rarity and conservation is the idea that rare species have a greater likelihood of extinction than do others (the terms rare, threatened and endangered are often used almost interchangeably; Table 1.7 provides their IUCN definitions). The significance of this observation to some degree rests upon what are regarded as the ultimate objectives of conservation. These may be varied. Examples include the preservation of individual species, and the maintenance of vulnerable environments, of representative examples of different ecosystems, or of those processes which are essential to the existence of humankind. While all might be argued to have a role to play, one or other has tended to take precedence during different periods in the history of the conservation movement. The development of an overall strategy in which each goal is given appropriate emphasis and in which conflicts between those who recognize different goals are minimized, remains one of the great challenges to this movement.

The conservation of rare species is tied foremost to the view that a central objective of conservation is the prevention or limitation of the extinction of species. It may equally be argued, of course, that goals other than the preservation of individual species necessitate this preservation. In some instances this will be true; in others it will not. The maintenance of ecosystem function, for example, may not necessitate the presence of many species. Because typically we do not understand the role played by the majority of species the preservation of them all can perhaps be justified on the precautionary principle. However, given our present knowledge of ecosystems, and the constraints on what can in practice be achieved, this seems likely to be at best a weak line of argument (Lawton, 1991).

For the purposes of this chapter, we shall by and large ignore the particular grounds upon which the prevention of extinction is deemed desirable. Rather, emphasis will be laid upon: how good an indication of the risk of extinction the designation of a species as rare provides; and the role of rarity in conservation. It should be noted that, as in the previous chapter, rarity will be used in the broad sense of a low abundance and/or a small range. This is different from the more constrained sense in which it is used in many compilations of the species whose survival is most threatened in an area (Chapter 1, and see below).

A central tenet of what follows will be that conservation necessitates prioritization. Whether of sites or of species, prioritization is plainly at odds with some ethical and moral standpoints, implying as it does that some sites and some species can be regarded as being of more value than others. It is shamelessly a pragmatic approach. We cannot hope to maintain all sites in their present condition, return them to some past state, or manage them in some other way which we feel to be optimal for the purposes of conservation. Likewise, we will not be able to prevent substantial reductions in the abundances or ranges, and ultimately the extinction of many, very probably a great many, species as a direct or indirect product of man's activities. There is insufficient time, insufficient funding and insufficient political will.

Prioritization, either of sites or of species, may be performed at a variety of spatial scales. Thus, for example, the managers of individual reserves, regional, national and international organizations, are all frequently concerned, at their own scales of interest, both with identifying those areas of prime conservation importance and those species which are most threatened. Priorities are apt to form a nested hierarchy in which sites and species recognized as important on a large spatial scale are likely to be regarded as also being important on a smaller scale, but not necessarily the converse. There are balances to be struck regarding the resources provided for conserving the priority areas and species recognized at different scales. While a species which is rare within a given district or county may be widespread and abundant internationally, its conservation within that area may be important to the local people and action to conserve it may well affect their commitment to wider conservation issues.

7.1   RARITY AND LIKELIHOOD OF EXTINCTION

Rarity, whether expressed in terms of abundance or range size, is undoubtedly a major determinant of a species' risk of extinction at the scale at which it was recognized as rare (Chapter 5). In this sense it provides a reasonable basis for identifying those species most in need of conservation at this scale. However, our ability to predict confidently the likelihood of a particular species becoming extinct is limited. This is because there remains a great deal of variation in this likelihood, which differences in abundance or range size cannot explain.

Table 7.1 Possible factors contributing to the extinction of local populations. (From Soulé, 1983.)

Rarity (low density)
Rarity (small, infrequent patches)
Limited dispersal ability
Inbreeding
Loss of heterozygosity
Founder effects
Hybridization
Successional loss of habitat
Environmental variation
Long-term environmental trends
Catastrophe
Extinction or reduction of mutualistic populations
Competition
Predation
Disease
Hunting and collecting
Habitat disturbance
Habitat destruction

This is not unexpected and a variety of other parameters are also likely to affect extinction probabilities (Table 7.1).

7.1.1 Species attributes

Even within a taxon, species differ in an array of traits which may alter their vulnerability to extinction. Those that have been postulated to do so include body size, habitat or diet specificity, longevity, dispersal ability and trophic level (Leck, 1979; Diamond, 1984b; Pimm *et al.*, 1988; Burbidge and McKenzie, 1989; Kattan, 1992; Laurance, 1991). However, the interrelationships of these traits, both with one another and with abundance and range size (Chapter 6) complicate interpretation of their individual importance. The example of body size in animals will suffice. Large-bodied species can have characteristics which render individuals less susceptible to vagaries in the environment than small species (Wasserman and Mitter, 1978 and references therein; Cawthorne and Marchant, 1980), and they also have abundances which are frequently, though not always, lower than those of small species (Chapter 6). The former may reduce the risk of extinction that large species face, while the latter may increase it. Indeed, there are studies which have failed to find an effect of body size on a species' risk of extinction, and studies which have reported positive or negative effects (Terborgh and Winter, 1980; Karr, 1982a; Diamond, 1984b; Pimm *et al.*, 1988; Soulé *et al.*, 1988; Burbidge and McKenzie, 1989; Gotelli and Graves, 1990; Maurer *et al.*, 1991; Kattan, 1992; Laurance, 1991; Tracy and George, 1992).

In addition to the very broad potential correlates of extinction probability, individual rare species may possess various characteristics which reduce their vulnerability to extinction below that which might have been predicted for organisms of their abundance or range size. Rabinowitz *et al.* (1989) demonstrate that sparse species of prairie grasses tend to have temporally less variable reproductive output than common species in the same habitat. This buffered output is achieved through growth and flowering during a season when rainfall is more predictable, and may compensate for one of the hazards of small population size – demographic stochasticity – and thus reduce the risk of local extinction. Such adaptations may variously be viewed as the results of selective pressure on individual organisms to enhance the survival probability of their own descendants, or as resulting from the differential survival of species which possess them (selective extinction).

7.1.2 Environmental attributes

As well as abundance, range size and other species characteristics, risks of extinction also depend on the attributes of the environment (Diamond, 1984a; Tracy and George, 1992). These might include temporal and spatial variation in habitat variables, and the frequency of catastrophes (infrequent but severe environmental perturbations, the probability of a population surviving through which is, at best, only weakly a function of its size). Again, the relationships between rarity, environmental variables and extinction probability need to be carefully dissected. Thus, the idea that species in tropical regions are more prone to extinction could be explained as a consequence of their adaptation to environments which, broadly speaking, are less temporally variable (Stevens, 1989) which makes them more vulnerable to severe perturbations when they do occur. However, it might equally be explained as a product of their tendency toward having, on average, smaller range sizes and lower densities.

The relative importance of species' characteristics and environmental attributes when determining the likelihood of extinction remains unclear. Indeed there has been little empirical exploration of the effects of environmental variables on extinction rates, beyond the effects of area and of isolation.

7.1.3 Population dynamics

Population dynamics are only defensible as a category separate from the previous two on the grounds of convenience, as they can be explained in terms of some combination of species' characteristics and environmental parameters. Some of the possible effects of a species' population dynamics on its likelihood of extinction have already been alluded to (Chapter 5).

The most obvious population dynamic clue to a species' risk of extinction is, of course, its temporal trajectory. Populations and range sizes in decline face inevitable extinction unless that trend can be altered. The absence of a decline does not, however, mean that a species is not at severe risk of extinction.

Concern has grown over a far more subtle population dynamic effect on probability of extinction. That is the distinction between sink and source populations (Pulliam, 1988). Local source populations in which reproduction exceeds mortality may sustain sink populations in which local reproduction fails to compensate for mortality. The sinks may, however, comprise a large proportion of the regional population (Pulliam, 1988; Howe *et al.*, 1991). Species whose regional populations are so structured will be at greater risk of extinction than will those in which a greater proportion of individuals occur in self-sustaining populations.

As noted earlier, the fact that rarity alone is not a sufficient predictor of the probability of extinction of a species is one of the reasons Munton (1987) provides for dissatisfaction over use of a rare category in some schemes by which species are classified as under differing degrees of threat (Chapter 1). A more acceptable approach is to categorize species under several different variables (discontinuous) which contribute to risk of extinction, and then to rank the various combinations of states to provide a sequence of conservation priorities. Indeed, as Munton (1987) acknowledges, this has been done in some schemes. Using such an approach necessitates decisions being made regarding the optimal choice of variables and the ranking of combinations of their values to ensure that the species most at risk receive greatest priority in conservation planning.

7.2   MORE LIMITATIONS OF RARITY

Although more refined methodologies are plainly desirable, information on the threat faced by many species is so poor that whether they are rare or not is the best indicator available of their need for active conservation efforts.

Unfortunately, for the majority of species, data are inadequate to achieve even a categorization as coarse as whether they are rare or not. In particular, this applies almost uniformly to invertebrate groups. The primary reason is the sheer magnitude of the task. The invertebrate fauna of Britain, which is widely acknowledged to be the best documented in the world, comprises some 29 000 known species, with several thousand more probably remaining to be found or recognized (Stubbs, 1982). Information on abundance and spatial occurrence suitable for making conservation decisions is available for a few thousand of the total at most. Obtaining such information for the greater proportion of species is an unrealistic objective (Disney, 1986). In order to maximize the conservation of wholesale biodiversity (a large part of overall biodiversity; Williams and Gaston, 1994) priorities will of necessity be based on the identification and conservation of vulnerable assemblages of species and not upon species by species' categorization schemes.

It should be emphasized that we have no idea what the composition might be of the assemblage of species distinguished across all taxa as being rare at the global scale. Nor do we know the composition of the assemblage of, say, the 10% of species which are most threatened with extinction at the present time. A number of taxa have claims to disproportionate representation in the rare category when compared with their total species' richness. It has, for example, been argued that plant species are more prone to persistence at very small global population sizes than are other higher taxa and that most insect species have very small geographic ranges. A useful way forward might be to generate estimates of the breadth of values for the geographic range sizes of different groups of species. A few such figures are already available. Thus, Solem (1984) predicts that the median geographic range size of all land snail species will be considerably less than 100 km, and probably less than 50 km. Bibby *et al.* (1992b) report that 2609 landbird species have geographic ranges of 50 000 km^2 or less (27% of all bird species). One suspects that further work of this kind might reveal that the globally rare species comprise a fascinating, and rather surprising, mixture of taxa.

The biases in lists of rare, threatened or endangered species should lead to some caution in their translation into a set of conservation priorities. This translation might additionally be modified by the levels of protection which species are already experiencing, or the ease with which protection can be undertaken. McIntyre (1992) stresses the need to recognize groups of species whose conservation requirements have been neglected or underestimated.

7.3   SITES AND SPECIES

If we accept that the conservation of rare species is, for whatever reason, a worthy goal, then how in practice can it be achieved? Although the division is probably more artificial than it at first seems, two different strategies can be recognized, one site or habitat oriented, the other oriented about individual species.

7.3.1 Site prioritization

The majority of species will be conserved not through direct management of their populations, but, more indirectly, through the preservation of the habitats or sites (defined broadly in this context to include areas of any size) in which they occur. A variety of criteria have been proposed by which the most important sites for conservation can be identified. These include diversity (in its various meanings), rarity, numbers of biological interactions (e.g. predatory, competitive), representativeness (how unique or typical sites are), naturalness, ecological fragility, area, degree of threat, scientific value, potential value, management potential, spatial position, replaceability, amenity value, recorded history, educational value, and ease of acquisition (for reviews see Margules and Usher, 1981; Spellerberg, 1981, 1992; Goldsmith, 1983, 1991b; Usher, 1986a). Different studies have used varying subsets of this list, although some criteria are commonly applied (see Margules and Usher, 1981; Usher, 1986a).

(a) Rarity as a criterion

The presence of rare species has been used as a criterion by which to prioritize areas in many studies (e.g. Gehlbach, 1975; Goldsmith, 1975, 1987; Wright,

1977; van der Ploeg and Vlijm, 1978; Fuller, 1980; Game and Peterken, 1984; Dony and Denholm, 1985; Nilsson, 1986; Slater *et al.*, 1987; Wheeler, 1988; Eyre and Rushton, 1989; Daniels *et al.*, 1991). Indeed, rarity ranks as one of the most frequently applied criteria. Primarily this doubtless reflects the importance placed on the conservation of rare species. It has also been suggested that rarity is one of the more readily quantified criteria (Goldsmith, 1983), however, this is only because the bulk of studies of prioritization have used data for vertebrate or plant taxa in temperate regions.

No general standardized methodologies have been arrived at by which rarity can be scored for the purposes of site prioritization. Rather, *ad hoc* methods tend to be developed dependent upon the abundance and/or range size data available for any particular study. Rarity scores can simply be based on the number of rare species occurring at a site, or calculated as some product of the levels of rarity achieved by each species at the site (discontinuous versus continuous measures). By and large the latter approach tends to be used, providing as it does both greater information content and a more refined basis for decision making.

While for the purposes of site prioritization rarity has been measured on the basis of both the range sizes and abundances of species, range sizes are most frequently used, as estimates of their magnitude are more often available (exceptions include the study of Fuller (1980) who had access to abundance data for British birds). If abundances are used, range sizes are usually used as well. A potentially important, but generally overlooked, consequence of primarily using the range sizes of species rather than species' abundances for prioritization is that comparatively little weight is given to species which occur widely at very low densities. The populations of such species may need to be conserved at many sites if they are to remain viable. It is notable that several of the families of landbirds found to have no species with ranges less than 50 000 km^2, and therefore not contributing to the scheme of priority areas determined by Bibby *et al.* (1992b), comprised large-bodied species (e.g. bustards, storks, cranes) which often occur at very low densities.

Site prioritization based, if only in part, on rarity, only makes sense if species are categorized as rare or otherwise at some higher spatial scale than that of the sites themselves. Species are thus typically categorized as rare or otherwise with respect to their spatial distribution or abundance across the geographic area within which all the sites occur. Categorizing species with respect to the individual site would be unhelpful because species which are rare with respect to the whole geographic area, and therefore most in need of conservation, may be abundant at one or more individual sites. Likewise, species which are rare with respect to individual sites may be abundant with respect to the whole geographic region.

Which higher spatial scale is chosen for defining rarity may, of course, affect where different component areas lie in the resultant sequence of priorities. One way of trying to account for this complication is to weight species on the basis of their rarity at more than one spatial scale (e.g. Jefferson and Usher, 1986;

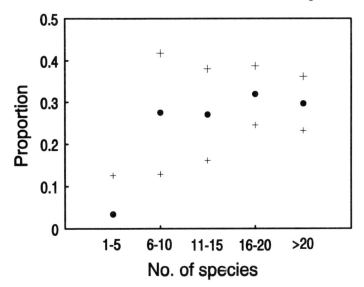

Figure 7.1 The proportion of species and subspecies which had restricted geographic ranges in butterfly assemblages of differing richness in the Moroccan Atlas mountains (restricted range species are those with at least one third of their range in North Africa, and restricted range sub-species are those with at least half of their range in North Africa). Dots are means, and crosses ± 1 SD. (Redrawn from Thomas and Mallorie, 1985b.)

Rapoport *et al.*, 1986; van der Ploeg, 1986; Daniels *et al.*, 1991), or to weight them according to the scale at which they are rare (e.g. Brooker and Welsh, 1982; Jenkins *et al.*, 1984). Thus, for example, in their study of the conservation value of ecological zones, habitat types and specific localities in a south Indian district, Daniels *et al.* (1991) first assigned conservation values to each of the bird species of the district. These values were based on seven variables, four of which related to the species' geographic ranges at different scales (over the entire world, over the Oriental region, over the Indian subcontinent and over the Malabar province).

It is commonly observed that, using a variety of definitions of rarity, the number of rare species at a site tends to be positively correlated with the overall species richness (Figure 7.1; Pearson, 1977; Järvinen, 1982; White *et al.*, 1984; Thomas and Mallorie, 1985b; C. Nilsson *et al.*, 1988; Wheeler, 1988; Dzwonko and Loster, 1989). It has also been found that the mean range size of species in an assemblage declines with the numbers of species present, which may amount to much the same thing (Figure 7.2; Rosenzweig, 1975; McCoy and Connor, 1980; Anderson and Koopman, 1981; Anderson, 1984a, b, 1985). In prioritization studies these observations translate into a tendency for high rarity scores to be associated with greater species' richness, either because rarity scores are simply the numbers of rare species or because they are correlated with these numbers.

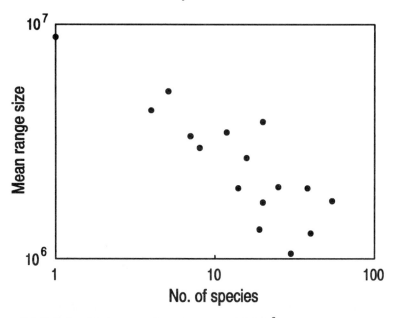

Figure 7.2 Relationship between the mean range size (km^2) of the species of reptiles occurring at different sample sites in North America and the number of species occurring at those sites. Some data points refer to more than one site. (Redrawn from Anderson, 1984b.)

Any tendency on the part of rare species to aggregate may be welcome, as it potentially reduces the numbers of sites that need to be protected in order to conserve them. Indeed, from this perspective the optimal scenario is perhaps that in which the ranges of species in an assemblage are nested such that species only occur in those places where the species with the next largest range also occur. In terms of the potential ease with which large numbers of species could be conserved it is encouraging to find that some 20% of all bird species are confined to just 2% of the earth's land area (Bibby *et al.*, 1992b). However, equally this emphasizes how essential it is to protect those areas.

As yet, our abilities to predict the numbers of rare species of a given taxon in an area are limited. Documented relationships between environmental characteristics (e.g. topography, habitat, climate) and the richness of rare species are increasingly being seen as a way round this problem, on the rationale that it is far easier to obtain data on the environmental conditions in an area than to document its fauna or flora adequately. Relationships between environmental characteristics and known patterns of occurrence have been explored in a number of studies (White *et al.*, 1984; Miller, 1986; Miller *et al.*, 1987, 1989; C. Nilsson *et al.*, 1988; White and Miller, 1988; Hill and Keddy, 1992). Hill and Keddy (1992), for example, developed predictive models relating the species' richness of rare plants to measured habitat variables for the shoreline vegetation of lakes in southwestern Nova Scotia. Multiple regression models using habitat variables accounted for 83% of the variability in species' richness of rare coastal plain species, compared with only 45% of that for the background flora. More studies of this kind are needed, as is some consideration of whether there are any generalizations to be made as to the combinations of variables which have the best ratio of predictive value to obtainability.

(b) Weighting different criteria

Alone, the rarity scores of sites will often provide an insufficient basis for prioritization. As noted earlier, rarity is one of many criteria upon which sites can be prioritized. Even if the sole intention were the preservation of the rare species themselves, this might most effectively be achieved at sites which do not rank highest on the rarity scoring, but on other, or a combination of other, variables. Paradoxically, as Goldsmith (1987) points out, if judged by rarity alone, the conservation value of a site could be increased by making its rarities even rarer rather than by reducing their vulnerability to extinction.

In most studies, a final ranking of sites is generated by combining the scores of different criteria to generate a single index. Typically no differential weighting is applied to the various scores, largely because there is no obvious way to do so. Indices can either be calculated once or iteratively, in the latter case on the assumption that, for example, at each step the site with highest scores has been conserved (e.g. Kirkpatrick, 1983). Iterative models are more efficient than simple scoring approaches, in that they can, for instance, more effectively ensure maximization of conservation criteria in the fewest or smallest areas (Margules *et al.*, 1988; Pressey and Nicholls, 1989a, b).

(c) Taxa

It has often been assumed, albeit tacitly, that the optimal ranking of sites for the conservation of one taxonomic group of species (usually plants) will also be optimal for other taxa occurring at those same sites. Such an assumption may equally often be false. Although not of itself a sufficient condition, this is implied by the observation that the patterns of species' richness of different taxa are often far from congruent, both at large (e.g. Schall and Pianka, 1978; Pianka and Schall, 1981; MacKinnon and MacKinnon, 1986b; Currie, 1991; Gentry, 1992) and at small spatial scales (Emberson, 1985; Usher, 1986b). More directly, Ryti (1992) has demonstrated that the use of different taxa as the basis for the selection of nature reserves can profoundly alter success at the inclusion of other taxa.

This is not to say that the richness of one taxon never provides information on the richness of another. There are ample examples of relationships between the numbers of insect species and the numbers of plant species in different areas (e.g. Abbott, 1974; Otte, 1976; Hockin, 1981; Reed, 1982; Itamies, 1983; S.G. Nilsson *et al.*, 1988; Brown and Opler, 1990). Even where such relationships exist independently of area effects, they need to be treated cautiously, in that one or other taxon may increase in richness at a rate that is disproportionate compared with the other. Thus, areas which differ markedly in the richness of one taxon may not differ so markedly in the richness of another.

There is, however, for most regions of the world little alternative to basing the prioritization of sites on one taxon or a few taxa. The extent to which this strategy serves to conserve species of other taxa will have a profound effect on how future generations view the success of present-day conservation efforts.

(d) Design, acquisition and dynamics

There are inadequacies in the existing reserve system. Reserves cover an insufficient proportion of the world's land surface (at present about 3–4%, and far less for marine habitats), are haphazard in distribution, tend to be located in areas of no economic value rather than in areas of high importance for conservation, and are often (usually?) too small (Spellerberg, 1991). If this situation is to be rationally improved in an optimal fashion, it will necessitate a closer marriage between the procedures by which sites are prioritized and many of the principles of good reserve design and management (see Speller-berg (1991) and Morris (1991) and references therein for entries into the literature on these topics).

Some aspects of these principles are included reasonably frequently among the criteria for prioritization studies (e.g. site area), but others are more seldom considered (e.g. site shape, location and connectivity; but see Bedward et al., 1992). In particular there is a general need to include information on the probability that conserved areas will remain conserved and will not be serious-ly degraded and on other social and political dimensions (see Soulé, 1991). This would go some way to answering critics of the derivation of rules for priori-tizing sites who argue that it ignores the practical reality that there is often no opportunity for their application.

Schemes of prioritization will also have to address two further, linked, issues. The first is a recognition of the need to move further towards a matrix approach to conservation, based not only on reserves, but also on non-reserve lands, often non-pristine lands. It is increasingly understood that a successful conservation strategy must include planning for the landscape as a whole. There are several reasons. Reserves themselves do not exist independently of the areas that surround them, reserve systems alone will never be adequate, and non-reserve lands frequently provide important areas of habitat. By way of an extreme example of the final point, comparison of the floras of seven English counties revealed that industrial sites are the sole or major habitat for between 9 and 31% of the species occurring in those counties (Kelcey, 1984).

The second issue that will need to be addressed is the fundamental flaw in a conservation strategy that focuses primarily upon a static system of isolated reserves. It ignores the fact that the geographic ranges of species shift in

Table 7.2 Means by which populations of rare species can be increased or maintained

Establishment of protected areas
Captive breeding
Supplemental feeding
Habitat manipulation
Reintroduction (from captive or wild populations) into areas where previously did occur
Introduction (from captive or wild populations) into areas where previously did not occur
Translocation between areas where already occur (e.g. to facilitate maintenance of genetic diversity)
Elimination or reduction of competitors, predators or parasites
Abolition of exploitation (e.g. hunting, collecting)
Manipulation of reproductive biology
Manipulation of behaviour (e.g. fencing to restrict dispersal)
Veterinary, sylvicultural and equivalent care of wild individuals

response to environmental change (Chapter 5; Huntley, 1991). At the least, it is necessary to have a network of interconnected sites. Such considerations are of particularly great importance in the light of the rapid changes in climate predicted in coming decades.

7.3.2 Focal species

The distinction between an approach to the conservation of rare species which is centred on habitats or sites, and one that is centred on particular species is blurred because focal (or 'flagship') species can only be preserved in the wild through adequate preservation of habitats. The latter inevitably leads to the preservation of non-focal species. The difference in the numbers of rare species which will be conserved effectively by the two approaches is liable to be case dependent, and a function of the degree to which general, good reserve design principles are practised.

This said, species-by-species conservation is in practice usually an exercise in crisis control (Guerrant, 1992). It is quite simply the only means by which a growing number of species can be prevented from becoming extinct. The classification of species according to the degree to which they are threatened with extinction provides a means of recognizing crises, or perhaps more realistically at the present time, the most serious crises.

As in considering a site-based approach to conservation, I have no intention of providing any detailed prescription of the process of conserving given species. The means by which populations of rare species can be managed are as diverse as the possible causes of their rarity (Tables 7.2 and 7.3). Rather, I have sought to identify some of the principal issues involved in species-by-species conservation of rare species:

(i)   Ideally, the future survival of a particular rare species is achieved

Table 7.3 Suggested management actions to increase the likelihood of the survival of Hawaii's endangered birds. A, legal protection of natural habitats; B, elimination of exotics; C, physical restoration of habitats; D, intensive manipulation of birds; E, translocation of birds; F, captive propagation and release to the wild. (From Scott and Kepler, 1985.)

	A	B	C	D	E	F
Hawaiian Goose		+				
Hawaiian Duck	+					
Laysan Duck		+				
Hawaiian Hawk	+					
Common Moorhen	+					
American Coot	+					
Black-necked Stilt	+					
Hawaiian Crow	+					
Nihoa Millerbird					+	
Hawaiian Thrush		+				
Small Kauai Thrush		+				
Kauai Oo		+				
Laysan Finch					+	
Ou	+					
Palila	+					
Maui Parrotbill		+				
Common Amakihi		+				
Kauai Akialoa		+				
Nukupuu		+				
Akiapolaau	+					
Hawaii Creeper	+					
Molokai Creeper	+					
Oahu Creeper	+					
Akepa	+					
Crested Honeycreeper	+					
Poouli		+				

through recognition of the factors which limit its abundance and/or range size, followed by action to relieve those (or other) restrictions. The natural causes of rarity are multifarious and complex (Chapter 6), thus there are equally endless ways in which the future survival of a species can be reduced through human actions. In broad terms, of course, it is not necessary to look far to identify the major threats. The single most important threat is the ever changing pattern of land use; the seemingly remorseless destruction of natural habitats and their replacement with agricultural and urban developments, and the 'improvement' of much agricultural land (e.g. Tables 7.4 and 7.5). Land

Table 7.4 Causes and causal agents of the decline of threatened plant species in the former Federal Republic of Germany. Owing to multiple listing of species endangered by different factors the sum of species is greater than the total number of individual species. (From Sukopp and Trautmann, 1981.)

	Species
Causes (factors)	
Elimination of special habitats	210
Drainage	173
Abandonment of land use	172
Landfilling, grading	155
Alteration of land use	123
Strip mining, removal of top soil	112
Mechanical impacts (e.g. trampling, camping)	99
Herbicide application	89
Impacts of weeding, clearing, fire	81
Dredging of rivers and lakes	69
Collecting	67
Water eutrophication	56
Discontinuance of periodical soil disturbance	42
Water pollution	31
Urbanization of villages	20
Causal agents (land-use systems)	
Agriculture including land consolidation and improvement	397
Tourism and recreation	112
Mining of raw materials, quarrying	106
Urban – industrial utilization	99
Water management	92
Forestry and hunting	84
Disposal of refuse and waste water	67
Fish pond management	37
Military	32
Traffic and transportation	19
Science	7

use is the major factor determining commonness and rarity in many faunas and floras, as illustrated by Hodgson's (1986b, 1991) work on the British flora. Many, though perhaps not all, of the large scale declines of whole groups of species can be viewed as the results of habitat alteration (Holmes and Sherry, 1988; Terborgh 1989; van Swaay, 1990; Hill and Hagan, 1991; Desender and Turin, 1989).

(ii)   While general threats to rare species may often, though not always, be

Table 7.5 The numbers of taxa under different general
kinds of threat in the freshwater fish faunas of western
and eastern North America, north of Mexico. (From
Deacon, 1979.)

	Western		Eastern	
	N	(%)	N	(%)
Habitat modification	109	(97.3)	90	(100)
Overexploitation	0	(0)	6	(6.7)
Parasitism and disease	5	(4.4)	0	(0)
Biotic interactions	60	(54)	8	(8.9)
Restricted range	24	(21)	6	(6.7)

identifiable, the precise causal links whereby they bring about reductions in species' abundances or range sizes may be more difficult to recognize. There are a number of considerations:

● Ideally, the particular causes of the rarity of a given species are determined through detailed study of its biology. However, such studies are demanding in both time and resources, and are in most cases unrealistic undertakings. Much important knowledge can be gained through more cursory comparisons of some basic ecological and life history parameters (Hodgson, 1991), but even such data as these are often hard to come by.
● The causes of the rarity of a given species may not operate all of the time, but rather be expressed in brief 'crunch' periods. Detection of such effects may necessitate time series data.
● Determination of the causes of rarity of threatened species is, almost invariably, based on observation rather than experiment. This carries some inherent dangers. In particular, it is liable to lead to reliance on the presumed basis of documented correlative analyses. An example is the presumption that observed habitat and resource usage reflect the best available options (that is they are optimal). Gray and Craig (1991) make the point that such a conclusion should not be reached automatically for any species, let alone one facing extinction. The species may be able to do much better in habitats and with diets outside its pattern of observed usage. Indeed, in the past its ecological requirements may have looked somewhat different and its true flexibility may be greater than even past usage might imply.
    This observation serves to emphasize the potential applied importance of an issue considered earlier from a 'pure' perspective, namely the interactions between the realized and fundamental niche spaces of species and their realized abundances and ranges sizes (Chapter 6).
● Species can be too rare for the cause of their rarity to be determined. This was true of more than a third of the species in Hubbell and Foster's (1986) work on the woody plants of their study plot on Barro Colorado island (Table 7.6).

Table 7.6 Some causes of rarity among woody plant species with fewer than 50 individuals in a 50 ha plot on Barro Colorado Island. (From Hubbell and Foster, 1986.)

No. of species (%)		Apparent cause of rarity
9	(8.1)	Most plants too small for inclusion in census
12	(10.8)	Habitat restriction
42	(37.8)	Common in second-growth forest
5	(4.6)	Other (hybrid, selectively cut)
43	(38.7)	Too rare to determine

It is a particular problem when working on species which occur at very low densities in complex habitats.

(iii)   The price of failing to diagnose the detailed causes of species' rarity correctly can be high. Ignorance of the subtle habitat requirements of some British butterflies meant that the loss of many local populations could not be prevented until remedial action was taken (J.A. Thomas, 1991). Green and Hirons (1991) identify initial misdiagnosis of the causes of a species' plight as a cause of the often long (10 years or more) delay between the start of research on and the recovery of the populations of individual endangered species of birds.

Lande (1988) gives examples of two management plans, for the northern spotted owl (*Strix occidentalis caurina*) and the red-cockaded woodpecker (*Picoides borealis*), which were based primarily on population genetics, ignoring basic demographic factors, and as a result threatened the existence of the species they were designed to protect.

(iv)   There have been attempts to provide broad guidelines regarding the most appropriate management strategies for rare species, on the basis of the general causes of their rarity (e.g. they are relict species, regulated by biological factors, or isolates at the edge of their range; Main, 1984). There seems little prospect of such schemes proving of much practical value.

(v)   Harper (1981) relates that a straw poll of a lecture audience revealed that the majority did not wish to see rare species made more common. As he states, 'This would seem to indicate that by some remarkable chance the flora of 1980 is just right and that the conservationists' task is to maintain this condition!' Likewise, Kelcey (1984) notes the 'overwhelming desire to maintain a rare species as a rare species'. Attempting to maintain many rare species at present abundances is not, however, likely to be an efficient means of ensuring their continued survival. It will necessitate frequent intervention to prevent extinction, while not permitting the enhancement of their populations which would reduce this frequency. These comments highlight a broad question. At what point in a conservation programme is the abundance or range size of a species considered sufficient for the programme to be regarded as successful and terminated?

(vi)   Although seldom referred to, measures directed at the conservation of some species may threaten others. For example, improvements to the water supply for the Owens River pupfish (*Cyprinodon radiosus*) resulted in the near loss of one of two populations of a snail endemic to the Owens Valley in California (J.M. Scott *et al.*, 1987). Likewise, management for large mammals in the Addo Elephant National Park (eastern Cape province, South Africa) has reduced resource availability for the large, flightless, dung-feeding scarab (*Circellium bacchus*) which is restricted to the park and surrounding farms (Scholtz and Chown, 1993).

The potential for such clashes is greatest when an area is being managed for the benefit of a particular focal species, rather than when the reasons for management are more broadly based. They are also most likely to occur when the taxa concerned operate on very different spatial scales (e.g. they are of very different body size; Scholtz and Chown, 1993). This latter observation suggests that as more attention is paid to the conservation of invertebrates, and particularly insects, the incidence of documented cases of such problems is liable to increase.

Improved information on the distribution and vulnerability of rare species will serve to reduce the occasions on which the continued survival of one species is accidentally threatened through actions to preserve another. However, ultimately hard decisions may have to be made as to which species have priority. The emphasis presently laid upon the desirability of creating corridors by which habitat patches can be linked is one instance where such conflicts are likely to be acute. What is a corridor to one species is apt to be a barrier for another.

7.4   CONCLUDING REMARKS

History has already demonstrated humankind's inability (in practice rather than principle) to coexist with all the other species extant at any given time. As a direct result of human activities we have witnessed the loss of numerous species and suspect the loss of many more. It seems almost inevitable that unless conflicts between the resource demands of people and wildlife are not reduced drastically, mass extinction will result. Humans have, of necessity, to impact upon the world around them. Nonetheless, choices exist as to the form that impact takes. An understanding of rarity provides a tool whereby those areas and species likely to be the most important indicators of the success of those choices in permitting global coexistence of the maximal number of species can be judged.

8 Where next?

Most is likely to be gained when we walk to the edge of established knowledge and peer over.

P.A. Keddy (1989)

It seemed that the next minute they would discover a solution. Yet it was clear to both of them that the end was still far, far off, and that the hardest and most complicated part was only just beginning.

A.P. Chekhov (1943)

More than most, the concluding chapters of monographs are apt to be prejudiced and of somewhat limited durability. Notwithstanding, it remains a useful exercise to stand back from the details of earlier chapters, to take a broad view of the study of rarity and to highlight the directions which might most usefully be emphasized in future work. The three main sections of this chapter attempt to attain these ends. The first addresses the value of the continued explicit study of rarity. The second identifies some inadequacies in approaches to the study of rarity to date, emphasizing: (a) taxonomic bias, (b) the lack of comparative studies and their limitations, (c) the need to synthesize information from diverse literatures, (d) statistical complexities and (e) the role of experiments.

The third section is a personal shortlist of important issues in our understanding of rarity which remain to be resolved. These include: (a) the role of rare species, (b) the importance of adaptations to rarity, (c) the integration of dynamics across different scales, (d) the distinction between human and natural determinants of rarity and (e) the persistence of species with large ranges and low abundances.

8.1 WHY CONTINUE TO STUDY RARITY?

In the opening chapter some obvious criticisms of the study of rarity *per se* were touched on. Most significantly, rarity cannot be recognized as an entirely distinct phenomenon. There are, in general, no natural breaks in the spectrum of abundances or range sizes of the species comprising an assemblage. Would it therefore be more useful in future work to forget rarity as a useful study area in its own right? I believe that the answer, which is in essence a central thesis of this volume, lies in the old adage that to divide is to conquer. Almost all the study areas in population and community biology have fundamentally arbitrary limits. Where do populations, communities, competition, mutualism and

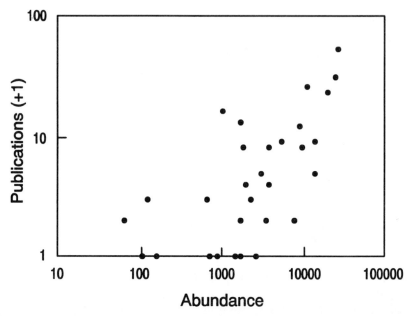

Figure 8.1 The relationship between the numbers of publications (listed in the Compact Cambridge Biological Index for 1982–1990) for species of North American sparrows and grosbeaks and the abundance of those species (as measured by the 1977 Breeding Bird Survey). (Redrawn from Kunin and Gaston, 1993.)

different guilds each end and begin? Study areas are delineated because they enable us to partition a research field in such a way as to produce a framework within which manageable questions can be conceived and addressed. For this reason alone we should continue to study rarity.

It could also be argued that failure in some quarters to recognize rarity as a topic of study has contributed to a strong bias in studies toward those species which are abundant and widespread (Figure 8.1). Moreover, it may have led to questions of more general importance to population and community biology being overlooked. Thus, while, for example, much attention has been focused on the question of whether species in an assemblage have stable relative abundances, the relative contributions of species of high and low abundances to observed patterns have largely been ignored (Chapter 5). Nonetheless, it remains essential to keep artificial divisions of research fields in proportion, '. . . without system the field of Nature would be a pathless wilderness; but system should be subservient to, not the main object of, pursuit' (Gilbert White, 1788).

Continuing to study rarity in its own right will have two primary consequences. First, it will enrich population and community biology as a whole. Indeed, there is a school which would argue that important generalities in ecology are more likely to emerge from comparative studies of common versus rare species, than from the more usual comparative studies of communities or assemblages (Bock, 1987). This follows from advocacy of an 'individualistic' approach to assemblage and community structure (Gleason, 1926). Second, continued study will improve the theoretical and practical foundation of conservation programmes, in that most threatened species are rare at the scale at which they are so designated.

8.2   HOW TO STUDY RARITY

In truth, there is no such thing as a recipe for the study of any subject area. Our knowledge of rarity has been enhanced by the diversity of approaches which have been taken. Nonetheless, it does seem that some aspects of the way in which rarity has been studied have been inadequately developed.

8.2.1 Taxonomic bias

Comments on the taxonomic biases that exist in studies of virtually all topics in population and community biology have often been made. Many would be equally relevant to work on rarity. While there has been some effort to draw examples for this volume from a diverse array of taxa, the vast majority of available data pertain to higher plants and birds, and to a lesser extent mammals. Moreover, by and large, three whole kingdoms have effectively been ignored; there has been little work on rarity in bacteria, viruses, protozoa and fungi.

This situation is broadly explicable, in terms of an innate attraction to work upon large and obvious taxa, and features which render their study more tractable. In the context of rarity, it in part also reflects the emphases of conservation. As well as the obvious desirability of enlarging the taxonomic base on which insights into rarity are founded, studies of rarity in additional taxa have further attractions. In particular, it seems likely that many experimental manipulations which are difficult to perform with large vertebrates would be more tractable with other groups.

In a related vein, it is also important that efforts are made towards standardizing the manner in which data are collected for different taxa. This would enable more rigorous between-taxa comparisons than are at present possible. This is particularly true with regard to measures of abundance and range size. Until further developments are made in this direction, answers to the question posed earlier as to what the globally rare species might be (Chapter 7) will remain elusive.

8.2.2 Comparative studies

'Surprisingly little is known about the overall biology of any individual plant species growing in the wild because scientists often choose plants to study in order to answer particular questions, not to find out all about them as organisms' (Holsinger and Gottlieb, 1991). Against this background, it is

Table 8.1 Various factors which may be important to a synthesis of the determinants of rarity in a given plant taxon. (From Fiedler, 1986.)

Age of taxon

a.   old and senescent, or
b.   young, incipient, or
c.   intermediate

Genotype of taxon

a.   depauperate/depleted, or
b.   heterogeneity comparable to closely related common species, or
c.   product of hybrid speciation

Evolutionary history

a.   effects of past climatic changes or stasis
b.   mode of origin of species

Taxonomic position

a.   meaningful taxonomic level (i.e. is rarity a taxonomic artefact?)
b.   systematic relationships – generic, specific, subspecific

Ecology

a.   habitat
b.   effects of present climate
c.   effects of edaphic conditions
d.   effects of predators/pathogens
e.   competitive ability

Population biology

a.   life history information
b.   status of populations
c.   factors influencing mortality and recruitment

Reproductive biology

a.   average number and range of flowers, fruits, seeds, seed set per reproductive individual
b.   pollination biology
c.   seed dispersal methods and/or agents, distance
d.   seed germination and establishment

Land use history

a.   habitat alteration
     1.   fire history – suppression?
     2.   introduction of exotic herbivores
     3.   introduction of exotic pollinators
     4.   introduction of exotic plant competitors
     5.   effects of current land management policies

Recent human uses

a.   horticultural trade
b.   aboriginal uses
c.   role in ancient and/or modern medicine and/or industry

---

perhaps unsurprising that there have been previous pleas for more comparative studies of the biologies of closely related rare and common species (Kruckeberg and Rabinowitz, 1985; Fiedler, 1987). The question of why the rare species in a given taxon have lower abundances and/or smaller range sizes than their relatives seems too broad to rest easily in the prevalent tradition of research on tightly focused hypotheses. This is particularly so because of the number and diversity of factors which may have to be integrated (Table 8.1). Inevitably studies have to become descriptive. Nonetheless, they need only be rudimentary to provide useful contributions to the search for patterns (or the demonstration of the lack of patterns) in the occurrence and causes of rarity. Most usefully, of course, they would comprise information on historical, biological, environmental and geographical characteristics of species (e.g. Fiedler, 1987), and involve both observational and experimental approaches.

8.2.3 Coping with phylogeny

To date, the phylogenetic background has largely been ignored in studies of rarity. It has, nonetheless, become apparent more broadly that the phylogenetic relatedness of species may have important effects on the outcomes of comparative studies (Harvey and Pagel, 1991). Where possible, appropriate methods of assessing the influence of phylogeny on observations should in future be applied.

Development of an understanding of the role of phylogeny in patterns of rarity will be hampered by continuing disagreement as to the most appropriate methods of coping with it (Eggleton and Vane-Wright, 1994) and, more importantly, the lack of sound phylogenies for the majority of groups. Obtaining sufficient series of specimens of many rare species to ascertain the full range of character variation they display may slow solutions to the latter problem.

8.2.4 The literature

An unfortunate consequence of the exponential historical increase in the number of published papers (Price, 1963) and the parallel increase in the specialism of individual researchers, has been a reduction in the exchange of ideas between closely related disciplines. A primary objective of this volume has been to synthesize work from several different areas of the biological literature. This can, however, only provide a start toward the process of breaking down the barriers generated by information overload.

Studies pertinent to an understanding of rarity can be found in the literatures concerned with botany, conservation, ecology, environmental management, genetics, palaeontology, statistics and zoology. In many instances these are not flagged as studies of rarity, indeed rarity and the various variants upon the term frequently fail to enter the indices of even major ecology texts. Further attempts to integrate information across these fields can serve only to enrich our understanding of rarity and provide fresh perspectives in the different subject areas. Perhaps the most obvious step to be taken in this direction would be an integration of genetic and population ecological insights. A foundation for this already exists in the burgeoning literature on the consequences of the genetic structure of populations for conservation (e.g. Schonewald-Cox *et al.*, 1983; Lande, 1988; Falk and Holsinger, 1991; Guerrant, 1992).

8.2.5 Statistical methods

Statistical complications to the study of rarity have been repeatedly highlighted in this book. Most notably, comparative studies are plagued by the difficulty of ensuring that observed assemblage or community patterns are not the products of sampling artefacts. In addition, it should be recognized that methods of analysing species-by-site data matrices are not influenced by rare species in identical ways. Thus, for example, rare species are known to contribute little to the magnitude of the Bray–Curtis similarity measure (Krebs, 1989), but greatly to other such measures.

Rare species are usually deleted from a data matrix prior to application of multivariate community-analysis methods of ordination and classification. Typical criteria for exclusion are that species occur in less than about 5%, or fewer than about 5–20 per hundred, of the samples (Gauch, 1982). The detailed justifications for such an approach may be varied. They include the observations that values for rare species (a) cause distortion of results (e.g. by being perceived as outliers), or increase the noise about underlying patterns and thereby obscure them, (b) do not contribute greatly to the information content of data matrices, (c) are not important components of communities (evidence is seldom offered in support of this; see below) and (d) increase the size of data matrices, necessitating greater computer storage facilities and escalating the computing time for analyses (difficulties which have substantially lessened with the availability of improved computing facilities).

The treatment of rare species should ultimately depend on the purposes of performing an analysis. As Faith and Norris (1989) point out, for example, the deletion of rare taxa in community pattern analysis may not be appropriate in the context of conservation. They found that the retention of rare taxa led to the recovery of environmental correlates of variation in the occurrence and abundance of macroinvertebrates which were not revealed by the analysis of common taxa alone. This had implications for the use of the representativeness of the ordination space as a criterion for reserve selection.

More broadly, Pianka (1986) expresses reluctance to drop rare species from analyses, on the grounds both that they exist and that they influence other species. Certainly, it seems desirable, wherever possible, both to pay attention to the means which do exist for reducing artefactual consequences of their inclusion in data matrices (e.g. Legendre and Legendre, 1983) and to develop statistical techniques which are more robust to the difficulties posed by rare species. Much might also be gained through concerted efforts to generate estimates of abundance and occurrence in which the probabilities of zeroes being real can be generated (McArdle, 1990) and through the development of techniques specifically to handle such data.

8.2.6 Experimental manipulations

The study of rarity has been dominated by observational studies. Experimental approaches have seldom been pursued. The primary reasons are doubtless the difficulties of manipulating rare species with sufficient replication, compounded in some instances with concerns either for the possible risks which manipulation might pose to their survival or that populations will not persist long enough for studies to be completed. While not wishing to minimize the reality of these problems, it seems that the benefits of a bolder approach might be substantial.

The rare species in many studies in population and community biology are not under any threat, and the problems of manipulation are almost exclusively those of obtaining adequate material with which to work. Much could almost certainly be achieved through long-term studies within which the level of replication of a manipulation (such as the transplantation of individuals in space) achievable at any one time may be low, but can in some sense be compensated for by repetition through time.

It seems reasonable to suppose that the more manipulations that are carried out with species which are rare to an area and yet not globally rare, the more expertise will become available for the manipulation of those species which are threatened with extinction. The growing realization that the observed habitat and resource use of particular rare species may not reflect either their flexibility or the conditions under which they do best (Gray and Craig, 1991) may make manipulation of these species an important management strategy.

8.3   SOME OUTSTANDING ISSUES

As elsewhere in population and community biology, it is easier to conceive of issues in the study of rarity which have yet to be resolved than of issues which have been satisfactorily investigated. From a host of possibilities of those which remain outstanding, I have chosen just a few which I consider to be of importance. The order in which they are presented implies nothing. Doubtless others would recognize different shortlists.

8.3.1 The role of rare species

In terms of the structure and maintenance of ecosystems, rare species have widely been regarded as having little significance. While one may feel ill at ease with this notion, it is not readily dismissed with hard data. Indeed, it seems quite plausible that systems could continue to operate in a substantially simplified condition, and it is not even clear that their evolutionary potential would necessarily be dramatically reduced by the removal of rare species. Pragmatism alone suggests two grounds for investigating the overall role played by rare species. First, many systems are inevitably going to be greatly simplified, and we need to predict the consequences. Second, if arguments that species richness (or biodiversity) is important for functional reasons are to be anything more than rhetoric, and ultimately to be widely seen as such, they need to be supported empirically.

Perhaps the most important question to answer is what proportion of rare species is ecologically extinct, 'ecological extinction' being the reduction of a species to such low abundance that although it is still present in a community it no longer interacts significantly with other species (Estes *et al.*, 1989). Put another way, how much redundancy is there in the biological composition of ecosystems (Walker, 1992)? It is of course very difficult to understand the function of many of the rare species in an ecosystem precisely because they are rare (Lovejoy, 1988). Moreover, it may be that the strength of the interactions of individual rare species is, of itself, comparatively insignificant, while the collective impact of many rare species on assemblage structure could be substantial (e.g. diffuse competition; MacArthur, 1972).

Some species which, with reference to the broad assemblages to which they belong, are rare over most spatial scales plainly have significant impacts on community structure. In particular, top predators may have profound effects through a variety of indirect channels (Terborgh, 1988). Their extirpation can lead to meso-predator release, with dramatic changes in species' richness and abundance patterns. Extinction of mega-herbivores may likewise precipitate extinctions of other herbivores (Owen-Smith, 1989). Equally, the interdependence (exploitative or mutualistic) of rare species (e.g. specialist pollinators on rare plant species) may mean that the removal of one rare species results in the loss of several others, with minimal effects on the essential structure of the ecosystems in which they are embedded.

8.3.2 Causes and adaptations

Great emphasis has, quite rightly, been placed on identifying the factors which cause species to be rare. Such information provides a foundation both for an understanding of community structure, and for management decisions to increase the abundances and spatial occurrence of the vulnerable and threatened. This work needs, however, to be balanced by a more concerted effort towards understanding the adaptations which rare species display, and which reduce their vulnerability to extinction.

Writers have commented on the difficulty in understanding how some species manage to persist at the exceedingly low densities or population sizes at which they are found (e.g. Pianka, 1986; Walter, 1990; Stacey and Taper, 1992). Craig (1991) cites largely isolated populations of Forbe's parakeet (*Cyanoramphus auriceps*) and black robin (*Petroica traversi*) on Little Mangere

Island, bellbirds (*Anthornis melanura*) on Tiritiri Matangi Island, and Campbell Island teal (*Anas aucklandica nesiotis*) on Dent Island, which have probably persisted with less than 20 pairs for a century or more. Moore (1991) provides a case of a small isolated population of the dragonfly (*Oxygastra curtisi*) which lasted for at least 140 years in southern England; very small isolated populations of acid water dragonflies also seem to be able to exist without replenishment for many years. The apparent survival of some very small populations for long periods contributes to the difficulty of predicting with any reliability the likelihood of any given species becoming extinct. They seem to provide exceptions to almost all broad correlates of extinction proneness. In many instances, such populations probably simply survive by chance. It seems likely, however, that in other cases they possess adaptations which substantially reduce their probability of extinction below that expected for species of, say, their body size or generation time. An obvious pitfall to be avoided in attempting to determine whether rare species exhibit characteristics which do reduce their likelihood of extinction, is the generation of a litany of just-so stories.

A largely unexplored avenue of research is to determine whether characteristics which are selected for because individuals of a species find themselves in isolated populations of low size or density may ultimately become barriers to population growth and range expansion if those constraints become relaxed. For example, there is some evidence that as a consequence of their becoming isolated, in populations of some butterfly species the dispersal ability of individuals declines rapidly (Dempster, 1991). If it cannot be so readily regained, this may result in the spatial occurrence of those species being limited by dispersal should conditions improve.

8.3.3 Large scales and small scales

Exploration of the consequences of differences in the scales, spatial and temporal, at which populations and communities are examined has had a major impact on our understanding of their structure (Dayton and Tegner, 1984; Giller and Gee, 1987; Levin, 1992). Many of the differences in the ways in which rarity is perceived can be traced to differences in the scales at which it has initially been defined. It is in part as a response to this situation that in this volume the concept of rarity has not been constrained to any particular scales. Rather, rare species are recognized to exist at virtually any combination of spatial and temporal scales, but need not have the same identities at these different scales.

The difficulty experienced in drawing together the material developed herein, suggests that concerted attempts at the integration of the dynamics of species' abundances and spatial occurrences over a variety of spatial and temporal scales would prove profitable. The most significant obstacle to such work is the lack of high quality occurrence data at a spectrum of spatial and temporal scales. A way forward in this field, at least for spatial scales, may be provided through developments in the use of techniques of estimating species' occurrences from existing data.

8.3.4 Human and natural determinants of rarity

The dominant causes of recent extinctions of species and increasingly of local populations are human activities. In this sense, practical interests in the likelihood of extinction of different species should centre on their relative resilience to the loss, fragmentation and disturbance of habitats. Empirical studies continue, however, to draw heavily on data concerned with the probably more natural extinctions of local populations or parts of populations of species on small offshore islands (e.g. Diamond, 1984b; Pimm *et al.*, 1988; Tracy and George, 1992). We need to understand the extent to which the characteristics of species that have reduced their likelihood of extinction over the bulk of their existence are also likely to do so with regard to human activities.

A distinction can be drawn between species which would be rare in the absence of human activity (here rarity is often independent of population dynamics and even of population trend; Ferrar, 1989), and those which are rare because of it (Cody, 1986). It seems probable that the former may be less prone to extinction as a consequence of their rarity than the latter, being more likely either to have evolved adaptations which reduce their vulnerability or being the survivors of selective extinctions because they possess such characteristics.

8.3.5 Low abundances and large range size

The existence of species with large ranges which occur nowhere at high densities is generally regarded as noteworthy (Good, 1948; Rabinowitz, 1981a; Rabinowitz *et al.*, 1986). Such species contravene the broad generalization that local abundance and range size are positively related (Chapter 3). The question of how they persist is an important one, and yet has only really received serious consideration in the context of prairie grasses (Rabinowitz and Rapp, 1981, 1985; Rabinowitz *et al.*, 1984, 1989). Here, aspects of competition, dispersal, establishment and reproductive output have been investigated. Further exploration of this topic for other groups should have high priority, because sparse species may be particularly vulnerable to erosion of the size of their geographic ranges. Their low population densities make them prone to local extinction and only by sustaining large geographic ranges are they liable to persist (Brown and Maurer, 1989).

8.4 CONCLUSION

There is something inherently paradoxical about work in population and community biology at the present time. On the one hand numerous supposed 'truths' as to the patterns and mechanisms of the natural world are being undermined. Increased awareness of statistical artefacts, and of the roles of scale and history have brought into question many previously accepted generalizations. This seems to be typical of critical periods in the progress of any science and heightened intensity of study is accompanied by an awareness of the paucity of knowledge.

On the other hand, practical answers are urgently needed to a wide range of questions, if the natural world is to be preserved for future generations in anything reminiscent of even its present state. The study of rarity provides a good example of the antagonism between these two positions.

References

Abbot, I. (1974) Number of plant, insect and landbird species on nineteen remote islands in the southern hemisphere. *Biological Journal of the Linnean Society*, **6**, 143–52.

Adams, D.E. and Anderson, R.C. (1982) An inverse relationship between dominance and habitat breadth of trees in Illinois forests. *American Midland Naturalist*, **107**, 192–5.

Adsersen, H. (1989) The rare plants of the Galapagos Islands and their conservation. *Biological Conservation*, **47**, 49–77.

Aizen, M.A. and Patterson, W. A. III (1990) Acorn size and geographical range in the North American oaks (*Quercus* L.). *Journal of Biogeography*, **17**, 327–32.

Anderson, S. (1977) Geographic ranges of North American terrestrial mammals. *American Museum Novitates*, **2629**, 1–15.

Anderson, S. (1984a) Geographic ranges of North American terrestrial birds. *American Museum Novitates*, **2785**, 1–17.

Anderson, S. (1984b) Areography of North American fishes, amphibians and reptiles. *American Museum Novitates*, **2802**, 1–16.

Anderson, S. (1985) The theory of range-size (RS) distributions. *American Museum Novitates*, **2833**, 1–20.

Anderson, S. and Koopman, K.F. (1981) Does interspecific competition limit the sizes of ranges of species? *American Museum Novitates*, **2716**, 1–10.

Andrewartha, H.G. and Birch, L.C. (1954) *The Distribution and Abundance of Animals*, University of Chicago Press, Chicago.

Angermeier, P.L. and Schlosser, I.J. (1989) Species-area relationships for stream fishes. *Ecology*, **70**, 1450–62.

Arita, H.T., Robinson, J.G. and Redford, K.H. (1990) Rarity in Neotropical forest mammals and its ecological correlates. *Conservation Biology*, **4**, 181–92.

Avery, M.I. and Haines-Young, R.H. (1990) Population estimates for the dunlin *Calidris alpina* derived from remotely sensed satellite imagery of the Flow Country of northern Scotland. *Nature*, **344**, 860–2.

Avise, J.C. (1992) Molecular population structure and the biogeographic history of a regional fauna: a case history with lessons for conservation biology. *Oikos*, **63**, 62–76.

Ayres, J.M. and Clutton-Brock, T.H. (1992) River boundaries and species range size in Amazonian primates. *American Naturalist*, **140**, 531–7.

Barnham, M. and Foggitt, G.T. (1987) *Butterflies in the Harrogate District*, privately published.

Barrett, S.C.H. and Kohn, J.R. (1991) Genetic and evolutionary consequences of small population size in plants: implications for conservation, in *Genetics and Conservation of Rare Plants* (eds D.A. Falk and K.E. Holsinger), Oxford University Press, Oxford, pp. 3–30.

Bart, J. and Klosiewski, S.P. (1989) Use of presence–absence to measure changes in avian density. *Journal of Wildlife Management*, **53**, 847–52.

Baskin, J.M. and Baskin, C.C. (1979) The ecological life cycle of the cedar glade endemic *Lobelia gattingerii. Bulletin of the Torrey Botanical Club*, **106**, 176–81.

Basset, Y. and Kitching, R.L. (1991) Species number, species abundance and body length of arboreal arthropods associated with an Australian rainforest tree. *Ecological Entomology*, **16**, 391–402.

Bawa, K.S. and Ashton, P.S. (1991) Conservation of rare trees in tropical rain forests: A genetic perspective, in *Genetics and Conservation of Rare Plants* (eds D.A. Falk and K.E. Holsinger), Oxford University Press, Oxford, pp. 62–71.

Beddington, J.R., Free, C.A. and Lawton, J.H. (1976) Concepts of stability and resilience in predator–prey models. *Journal of Animal Ecology*, **45**, 791–816.

Bedward, M., Pressey, R.L. and Keith, D.A. (1992) A new approach for selecting fully representative reserve networks: addressing efficiency, reserve design and land suitability with an iterative analysis. *Biological Conservation*, **62**, 115–25.

Beebe, W. (1925) Studies of a tropical jungle: one quarter of a square mile of jungle at Kartabo, British Guiana. *Zoologica*, **6**, 5–193.

Begon, M., Harper, J.L. and Townsend, C.R. (1990) *Ecology: Individuals, Populations and Communities* (2nd edn), Blackwell Scientific, Oxford.

Bibby, C.J., Burgess, N.D. and Hill, D.A. (1992a) *Bird Census Techniques*, Academic Press, London.

Bibby, C.J., Collar, N.J., Crosby, M.J. *et al.* (1992b) *Putting Biodiversity on the Map: Priority Areas for Global Conservation*, International Council for Bird Preservation, Cambridge.

Bibby, C.J. and Hill, D.A. (1987) Status of the Fuerteventura stonechat *Saxicola dacotiae. Ibis*, **129**, 491–8.

Blackburn, T.M., Brown, V.K., Doube, B. *et al.* (1993) The relationship between body size and abundance in natural animal assemblages. *Journal of Animal Ecology*, **62**, 519–28.

Blackburn, T.M., Harvey, P.H. and Pagel, M.D. (1990) Species number, population density and body size relationships in natural communities. *Journal of Animal Ecology*, **59**, 335–45.

Blakers, N., Davies, S.J.J.F. and Reilly, P.N. (1984) *The Atlas of Australian Birds*, Melbourne University Press, Carlton.

Bock, C.E. (1984) Geographical correlates of abundance vs. rarity in some North American winter landbirds. *The Auk*, **101**, 266–73.

Bock, C.E. (1987) Distribution–abundance relationships of some Arizona landbirds: a matter of scale? *Ecology*, **68**, 124–9.

Bock, C.E. and Ricklefs, R.E. (1983) Range size and local abundance of some North American songbirds: a positive correlation. *American Naturalist*, **122**, 295–9.

Bock, C.E. and Root, T.L. (1981) The Christmas Bird Count and avian ecology. *Studies in Avian Biology*, **6**, 17–23.

Bock, C.E., Bock, J.H. and Lepthien, L.W. (1977) Abundance patterns of some bird species wintering on the Great Plains of the U.S.A. *Journal of Biogeography*, **4**, 101–10.

Bowen, H.J.M. (1968) *The Flora of Berkshire*, Holywell Press, Oxford.

Bowers, M.A. (1988) Relationships between local distribution and geographic range of desert heteromyid rodents. *Oikos*, **53**, 303–8.

Brooker, M.P. and Welsh, W.A. (eds) (1982) *Conservation of Wildlife in River Corridors.*

Part 1. Methods of Survey and Classification, Welsh Water Authority, Cambrian Way, Brecon.

Brown, A.H.D. and Briggs, J.D. (1991) Sampling strategies for genetic variation in *ex situ* collections of endangered plant species, in *Genetics and Conservation of Rare Plants* (eds D.A. Falk and K.E. Holsinger), Oxford University Press, Oxford, pp. 99–119.

Brown, J.H. (1981) Two decades of homage to Santa Rosalia: toward a general theory of diversity. *American Zoologist*, **21**, 877–88.

Brown, J.H. (1984) On the relationship between abundance and distribution of species. *American Naturalist*, **124**, 255–79.

Brown, J.H. and Maurer, B.A. (1987) Evolution of species assemblages: effects of energetic constraints and species dynamics on diversification of the North American avifauna. *American Naturalist*, **130**, 1–17.

Brown, J.H. and Maurer, B.A. (1989) Macroecology: the division of food and space among species on continents. *Science (Wash.)*, **243**, 1145–50.

Brown, J.H. and Nicoletto, P.F. (1991) Spatial scaling of species composition: body masses of North American land mammals. *American Naturalist*, **138**, 1478–512.

Brown, J.W. and Opler, P.A. (1990) Patterns of butterfly species density in peninsula Florida. *Journal of Biogeography*, **17**, 615–22.

Browning, R. (1855) One Word More, in *Men and Women and Other Poems*, H. Chapman, London.

Brussard, P.F. (1984) Geographic patterns and environmental gradients: the central–marginal model: *Drosophila* revisited. *Annual Review of Ecology and Systematics*, **15**, 25–64.

Burbridge, A.A. and McKenzie, N.L. (1989) Patterns in the modern decline of Western Australia's vertebrate fauna: causes and conservation implications. *Biological Conservation*, **50**, 143–98.

Burgman, M.A. (1989) The habitat volumes of scarce and ubiquitous plants: a test of the model of environmental control. *American Naturalist*, **133**, 228–39.

Busby, J.R. (1986) A biogeoclimatic analysis of *Notofagus cunninghamii* (Hook.) Oestr. in southeastern Australia. *Australian Journal of Ecology*, **11**, 1–7.

Bushnell, J.H., Foster, S.Q. and Wahle, B.M. (1987) Annotated inventory of invertebrate populations of an alpine lake and stream chain in Colorado. *Great Basin Naturalist*, **47**, 500–11.

Buzas, M.A. and Culver, S.J. (1991) Species diversity and dispersal of benthic Foraminifera. *BioScience*, **41**, 483–9.

Buzas, M.A., Koch, C.F., Culver, S.J. and Sohl, N.F. (1982) On the distribution of species occurrence. *Paleobiology*, **8**, 143–50.

Cammell, M.E., Tatchell, G.M. and Woiwod, I.P. (1989) Spatial pattern of abundance of the black bean aphid, *Aphis fabae*, in Britain. *Journal of Applied Ecology*, **26**, 463–72.

Cappuccino, N. and Kareiva, P. (1985) Coping with a capricious environment: a population study of a rare pierid butterfly. *Ecology*, **66**, 152–61.

Carrascal, L.M. and Tellería, J.L. (1991) Bird size and density: a regional approach. *American Naturalist*, **138**, 777–84.

Carter, R.N. and Prince, S.D. (1985) The geographical distribution of prickly lettuce (*Lactuca serriola*). I. A general survey of its habitats and performance in Britain. *Journal of Ecology*, **73**, 27–38.

Carter, R.N. and Prince, S.D. (1988) Distribution limits from a demographic viewpoint, in *Plant Population Ecology* (eds A.J. Davy, M.J. Hutchings and A.R. Watkinson), Blackwell Scientific, Oxford, pp. 165–84.

Case, T.J. and Bolger, D.T. (1991) The role of introduced species in shaping the distribution and abundance of island reptiles. *Evolutionary Ecology*, **5**, 272–90.

Caughley, G. (1977) *Analysis of Vertebrate Populations*, Wiley, New York.

Cawthorne, R.A. and Marchant, J.H. (1980) The effects of the 1978/79 winter on British bird populations. *Bird Study*, **27**, 163–72.

Chekhov, A.P. (1943) *The Lady with the Toy Dog, and Gooseberries*, (Translated by S.S. Koteliansky), Todd, London.

Chesterton, G.K. (1909) *Orthodoxy*, John Lane, London.

Chiang, H.C. (1961) Fringe populations of the European corn borer, *Pyrausta nubilialis*: their characteristics and problems. *Annals of the Entomological Society of America*, **54**, 378–87.

Claridge, M.F. and Evans, H.F. (1990) Species–area relationships: relevance to pest problems of British trees? in *Population Dynamics of Forest Insects* (eds A.D. Watt, S.R. Leather, M.D. Hunter and N.A.C. Kidd), Intercept, Andover, pp. 59–69.

Claridge, M.F. and Wilson, M.R. (1981) Host plant associations, diversity and species–area relationships of mesophyll-feeding leafhoppers of trees and shrubs in Britain. *Ecological Entomology*, **6**, 217–38.

Claridge, M.F. and Wilson, M.R. (1982) Insect herbivore guilds and species–area relationships: leafminers on British trees. *Ecological Entomology*, **7**, 19–30.

Cody, M.L. (1986) Diversity, rarity, and conservation in Mediterranean-climate regions, in *Conservation Biology: the Science of Scarcity and Diversity* (ed. M.E. Soulé), Sinauer Associates, Sunderland, MA, pp. 122–52.

Colinvaux, P. (1980) *Why Big Fierce Animals Are Rare*, Penguin Books, Harmondsworth, UK.

Collar, N.J. and Stuart, S.N. (1985) *Threatened Birds of Africa and Related Islands: the ICBP/IUCN Red Data book, Part 1., ICBP and IUCN, Cambridge.*

Collins, S.L. and Glenn, S.M. (1990) A hierarchical analysis of species' abundance patterns in grassland vegetation. *American Naturalist*, **135**, 633–48.

Collins, S.L. and Glenn, S.M. (1991) Importance of spatial and temporal dynamics in species regional abundance and distribution. *Ecology*, **72**, 654–64.

Cornell, H.V. (1982) The notion of minimum distance or why rare species are clumped. *Oecologia (Berl.)*, **52**, 278–80.

Craig, J.L. (1991) Are small populations viable? *Acta XX Congressus Internationalis Ornithologici*, 2546–52.

Crowley, P.H. and Johnson, D.M. (1992) Variability and stability of a dragonfly assemblage. *Oecologia (Berl.)*, **90**, 260–9.

Currie, D.J. (1991) Energy and large-scale patterns of animal and plant-species richness. *American Naturalist*, **137**, 27–49.

Currie, D.J. (1993) What shape is the relationship between body size and population density? *Oikos*, **66**, 353–8.

Damuth, J. (1981) Population density and body size in mammals. *Nature*, **290**, 699–700.

Damuth, J. (1987) Interspecific allometry of population density in mammals and other animals: the independence of body mass and population energy-use. *Biological Journal of the Linnean Society*, **31**, 193–246.

Daniels, R.J.R., Hegde, M., Joshi, N.V. and Gadgil, M. (1991) Assigning conservation value: a case study from India. *Conservation Biology*, **5**, 464–75.

Darwin, C. (1859) *On The Origin of Species by Means of Natural Selection, or The Preservation of Favoured Races in the Struggle for Life*, John Murray, London.

Davis, M.B. (1976) Pleistocene biogeography of temperate deciduous forests. *Geoscience and Man*, **13**, 13–26.

Davis, N. and VanBlaricom, G.R. (1978) Spatial and temporal heterogeneity in a sand bottom epifaunal community of invertebrates in shallow water. *Limnology and Oceanography*, **23**, 417 –27.

Davis, S.D., Droop, S.J.M., Gregerson, P. *et al.* (1986) *Plants in Danger: What do we know?*, International Union for Conservation of Nature and Natural Resources, Cambridge, UK and Gland, Switzerland.

Dayton, P.K. and Tegner, M.J. (1984) The importance of scale in community ecology: A kelp forest example with terrestrial analogs, in *A New Ecology: Novel Approaches to Interactive Systems* (eds P.W. Price, C.N. Slobodchikoff and W.S. Gaud), Wiley, New York, pp. 457–81.

Deacon, J.E. (1979) Endangered and threatened fishes of the west. *Great Basin Naturalist Memoirs*, **3**, 14–64.

Dempster, J.P. (1991) Fragmentation, isolation and mobility of insect populations, in *The Conservation of Insects and their Habitats* (eds N.M. Collins and J.A. Thomas), Academic Press, London, pp. 143–53.

DeSante, D. and Pyle, P. (1986) *Distributional Checklist of North American Birds. Vol. 1. United States and Canada*, Artemisia Press, Lee Vining, CA.

Desender, K. and Turin, H. (1989) Loss of habitats and changes in the composition of the ground and tiger beetle fauna in four West European countries since 1950 (Coleoptera: Carabidae, Cicindelidae). *Biological Conservation*, **48**, 277–94.

Deshaye, J. and Morisset, P. (1989) Species–area relationships and the SLOSS effect in a Subarctic archipelago. *Biological Conservation*, **48**, 265–76.

Dhondt, A.A. (1988) Carrying capacity: a confusing concept. *Acta Oecologica*, **9**, 337–46.

Diamond, J.M. (1980) Patchy distributions of tropical birds, in *Conservation Biology: An Evolutionary–Ecological Perspective* (eds M.E. Soulé and B.A. Wilcox), Sinauer Associates, Sunderland, MA, pp. 57–74.

Diamond, J.M. (1984a) Historic extinctions: a Rosetta stone for understanding prehistoric extinctions, in *Quaternary Extinctions: A Prehistoric Revolution* (eds P.S. Main and R.G. Klein), University of Arizona Press, Tucson, AZ, pp. 824–62.

Diamond, J.M. (1984b) 'Normal' extinctions of isolated populations, in *Extinctions* (ed. M.H. Nitecki), University of Chicago Press, Chicago, IL, pp. 191–246.

Diamond, J.M. (1985) How many species are yet to be discovered? *Nature*, **315**, 538–9.

Digby, P.G.N. and Kempton, R.A. (1987) *Multivariate Analysis of Ecological Communities*, Chapman & Hall, London.

Disney, R.H.L. (1986) Assessments using invertebrates: posing the problem, in *Wildlife Conservation Evaluation* (ed. M.B. Usher), Chapman & Hall, London, pp. 271–93.

Disney, R.H.L. (1987) The use of rapid sample surveys of insect fauna, in *The Use of Invertebrates in Site Assessment for Conservation* (ed. M. Luff), Agricultural Environment Research Group, University of Newcastle Upon Tyne, pp. 19–26.

Dixon, A.F.G. and Kindlmann, P. (1990) Role of plant abundance in determining the abundance of herbivorous insects. *Oecologia (Berl.)*, **83**, 281–3.

Dony, J.G. (1953) *Flora of Bedfordshire*, The Corporation of Luton Museum and Art Gallery, Luton.

Dony, J.G. (1967) *Flora of Hertfordshire*, Hitchin Urban District Council, Hitchin.

Dony, J.G. and Denholm, I. (1985) Some quantitative methods of assessing the conservation value of ecologically similar sites. *Journal of Applied Ecology*, **22**, 229–38.

Doube, B.M. (1991) Dung beetles of southern Africa, in *Dung Beetle Ecology* (eds I. Hanski and Y. Cambefort), Princeton University Press, Princeton, NJ, pp. 133–55.

Drake, J.A., Mooney, H.A., di Castri, F. *et al.* (eds) (1989) *Biological Invasions: A Global Perspective*. Wiley, Chichester.

Dressler, R.L. (1982) *The Orchids: Natural History and Classification*, Harvard University Press, Cambridge, MA.

Dring, M.J. and Frost, L.C. (1971) Studies of *Ranunculus ophioglossifolius* in relation to its conservation at the Bridgenorth Nature Reserve, Gloucestershire, England. *Biological Conservation*, **4**, 48–56.

Drury, W.H. (1974) Rare species. *Biological Conservation*, **6**, 162–9.

Drury, W.H. (1980) Rare species of plants. *Rhodora*, **82**, 3–48.

Dueser, R.D. and Shugart, H.H. (1979) Niche pattern in a forest-floor small mammal fauna. *Ecology*, **60**, 108–18.

Dzwonko, Z. and Loster, S. (1989) Distribution of vascular plant species in small woodlands on the Western Carpathian foothills. *Oikos*, **56**, 77–86.

Ebeling, A.W., Holbrook, S.J. and Schmitt, R.J. (1990) Temporally concordant structure of a fish assemblage: bound or determined? *American Naturalist*, **135**, 63–73.

Eggleton, P. and Vane-Wright, R.I. (eds) (1994) *Phylogenetics and Ecology*, Academic Press, London.

Eisenberg, J.F., O'Connell, M.A. and August, P.V. (1979) Density, productivity, and distribution of mammals in two Venezuelan habitats, in *Vertebrate Ecology in the Northern Neotropics* (ed. J.F. Eisenberg), Smithsonian Institution, Washington, DC, pp. 187–207.

Elton, C. (1930) *Animal Ecology and Evolution*, Oxford University Press, New York.

Elton, C. (1932) Territory among wood ants (*Formica rufa* L.) at Picket Hill. *Journal of Animal Ecology*, **1**, 69–76.

Elton, C. (1933) *The Ecology of Animals,* Methuen, London.

Emberson, R.M. (1985) Comparisons of site conservation value using plant and soil arthropod species. *Bulletin of the British Ecological Society*, **16**, 16–17.

Emlen, J.T., DeJong, M.J., Jaeger, M.J. *et al.* (1986) Density trends and range boundary constraints of forest birds along a latitudinal gradient. *The Auk*, **103**, 791–803.

Erickson, R.O. (1945) The *Clematis fremontii* var. *riehlii* population in the Ozarks. *Annals of the Missouri Botanical Garden*, **32**, 413–60.

Espadaler, X. and López-Soria, L. (1991) Rareness of certain Mediterranean ant species: fact or artifact? *Insectes Sociaux*, **38**, 365–77.

Estes, J.A., Duggins, D.O. and Rathbun, G.B. (1989) The ecology of extinctions in kelp forest communities. *Conservation Biology*, **3**, 252–64.

Estes, J.A., Jameson, R.J. and Rhode, E.B. (1982) Activity and prey selection in the sea otter: influence of population status on community structure. *American Naturalist*, **120**, 242–58.

Eyre, M.D. and Rushton, S.P. (1989) Quantification of conservation criteria using invertebrates. *Journal of Applied Ecology*, **26**, 159–71.

Faith, D.P. and Norris, R.H. (1989) Correlation of environmental variables with patterns of distribution and abundance of common and rare freshwater macroinvertebrates. *Biological Conservation*, **50**, 77–98.

Falk, D.A. and Holsinger, K.E. (1991) *Genetics and Conservation of Rare Plants*, Oxford University Press, Oxford.

Falk, S. (1991) *A Review of the Scarce and Threatened Bees, Wasps and Ants of Great Britain*, Research and Survey in Nature Conservation 35, Nature Conservancy Council, Peterborough.

Felton, J.C. (1974) Some comments on the Aculeate fauna, in *The Changing Flora and Fauna of Britain* (ed. D.L. Hawksworth), Academic Press, London, pp. 399–418.

Fenchel, T. (1974) Intrinsic rate of increase: the relationship with body size. *Oecologia (Berl.)*, **14**, 317–26.

Ferrar, A.A. (1989) The role of Red Data Books in conserving biodiversity, in *Biotic Diversity in Southern Africa* (ed. B.J. Huntley), Oxford University Press, Cape Town, pp. 136–47.

Fiedler, P.L. (1986) Concepts of rarity in vascular plant species, with special reference to the genus *Calochortus* Pursh (Liliaceae). *Taxon*, **35**, 502–18.

Fiedler, P.L. (1987) Life history and population dynamics of rare and common mariposa lilies (*Calochortus* Pursh: Liliaceae). *Journal of Ecology*, **75**, 977–95.

Fiedler, P.L. and Ahouse, J.J. (1992) Hierarchies of cause: toward an understanding of rarity in vascular plant species, in *Conservation Biology: The Theory and Practice of Nature Conservation, Preservation and Management* (eds P.L. Fiedler and S.K. Jain), Chapman & Hall, London, pp. 23–47.

Fisher, R.A., Corbet, A.S. and Williams, C.B. (1943) The relation between the number of species and the number of individuals in a random sample of an animal population. *Journal of Animal Ecology*, **12**, 42–58.

Flesness, N.R. (1992) Living collections and biodiversity, in *Systematics, Ecology and the Biodiversity Crisis* (ed. N. Eldredge), Columbia University Press, New York, pp. 178–84.

Forcella, F. and Wood, J.T. (1984) Colonization potentials of alien weeds are related to their 'native' distributions: implications for plant quarantine. *Journal of the Australian Institute of Agricultural Science*, **50**, 35–40.

Forcella, F., Wood, J.T. and Dillon, S.P. (1986) Characteristics distinguishing invasive weeds within *Echium* (Bugloss). *Weed Research*, **26**, 351–64.

Ford, H.A. (1990) Relationships between distribution, abundance and foraging specialization in Australian landbirds. *Ornis Scandinavica*, **21**, 133–8.

Fowler, S.V. and Lawton, J.H. (1982) The effects of host–plant distribution and local abundance on the species richness of agromyzid flies attacking British umbellifers. *Ecological Entomology*, **7**, 257–65.

France, R. (1992) The North American latitudinal gradient in species richness and geographical range of freshwater crayfish and amphipods. *American Naturalist*, **139**, 342–54.

Francis, I.S., Penford, N., Gartshore, M.E. and Jaramillo, A. (1992) The White-breasted Guineafowl *Agelastes meleagrides* in Taï National Park, Côte d'Ivoire. *Bird Conservation International*, **2**, 25–60.

Frontier, S. (1987) Applications of fractal theory to ecology, in *Developments in Numerical Ecology* (eds P. Legendre and L. Legendre), Springer-Verlag, Berlin, pp. 335–78.

Fuller, R.J. (1980) A method for assessing the ornithological interest of sites for conservation. *Biological Conservation*, **17**, 229–39.

Game, M. and Peterken, G.F. (1984) Nature reserve selection strategies in the woodlands of central Lincolnshire, England. *Biological Conservation*, **29**, 157–82.

Gardner, S.M. (1991) Ground beetle (Coleoptera: Carabidae) communities on upland heath and their association with heathland flora. *Journal of Biogeography*, **18**, 281–9.

Gaston, K.J. (1988) Patterns in the local and regional dynamics of moth populations. *Oikos*, **53**, 49–59.

Gaston, K.J. (1991a) The magnitude of global insect species richness. *Conservation Biology*, **5**, 283–96.

Gaston, K.J. (1991b) How large is a species' geographic range? *Oikos*, **61**, 434–8.

Gaston, K.J., Blackburn, T.M. and Lawton, J.H. (1993) Comparing animals and automobiles: a vehicle for understanding body size and abundance relationships in species assemblages? *Oikos*, **66**, 172–9.

Gaston, K.J. and Lawton, J.H. (1988a) Patterns in body size, population dynamics and regional distribution of bracken herbivores. *American Naturalist*, **132**, 662–80.

Gaston, K.J. and Lawton, J.H. (1988b) Patterns in the distribution and abundance of insect populations. *Nature*, **331**, 709–12.

Gaston, K.J. and Lawton, J.H. (1989) Insect herbivores on bracken do not support the core-satellite hypothesis. *American Naturalist*, **134**, 761–77.

Gaston, K.J. and Lawton, J.H. (1990a) Effects of scale and habitat on the relationship between regional distribution and local abundance. *Oikos*, **58**, 329–35.

Gaston, K.J. and Lawton, J.H. (1990b) The population ecology of rare species. *Journal of Fish Biology*, **37** (Supplement A), 97–104.

Gaston, K.J. and McArdle, B.H. (1993) All else is not equal: temporal population variability and insect conservation, in *Perspectives on Insect Conservation* (eds K.J. Gaston, T.R. New and M.J. Samways), Intercept, Andover, in press.

Gauch, H.G., Jr. (1982) *Multivariate Analysis in Community Ecology*, Cambridge University Press, Cambridge.

Gehlbach, F.R. (1975) Investigation, evaluation and priority ranking of natural areas. *Biological Conservation*, **8**, 79–88.

Gentry, A.H. (1992) Tropical forest biodiversity: distributional patterns and their conservational significance. *Oikos*, **63**, 19–28.

Gerrard, D.J. and Chiang, H.C. (1970) Density estimation of corn rootworm egg populations based upon frequency of occurrence. *Ecology*, **51**, 237–45.

Gilbert, L.E. (1991) Biodiversity of a Central American *Heliconius* community: pattern, process, and problems, in *Plant–Animal Interactions: Evolutionary Ecology in Tropical and Temperate Regions* (eds P.W. Price, T.M. Lewinsohn, G.W. Fernandes and W.W. Benson), Wiley, New York, pp. 403–27.

Giller, P.S. and Gee, J.H.R. (1987) The analysis of community organisation: the influence of equilibrium, scale and terminology, in *Organisation of Communities: Past and Present* (eds J.H.R. Gee and P.S. Giller), Blackwell Scientific, Oxford, pp. 519–42.

Gilpin, M.E. and Hanski, I. (eds) (1991) *Metapopulation Dynamics*, Academic Press, London.

Gislen, T. and Kauri, H. (1959) Zoogeography of the Swedish amphibians and reptiles: with notes on their growth and ecology. *Acta Vertebrata*, **1**, 197–397.

Glazier, D.S. (1980) Ecological shifts and the evolution of geographically restricted species of North American *Peromyscus* (mice). *Journal of Biogeography*, **7**, 63–83.

Gleason, H.A. (1926) The individualistic concept of the plant association. *Bulletin of the Torrey Botanical Club*, **543**, 7–26.

Godfray, H.C.J. (1984) Patterns in the distribution of leafminers on British trees. *Ecological Entomology*, **9**, 163–8.

Goeden, R.D. and Ricker, D.W. (1986) Phytophagous insect faunas of the two most common native *Cirsium* thistles, *C. californicum* and *C. proteanum*, in southern California. *Annals of the Entomological Society of America*, **79**, 953–62.

Goldsmith, F.B. (1975) The evaluation of ecological resources in the countryside for conservation purposes. *Biological Conservation*, **8**, 89–96.

Goldsmith, F.B. (1983) Evaluating nature, in *Conservation in Perspective* (eds A. Warren and F.B. Goldsmith), Wiley, Chichester, pp. 233–46.

Goldsmith, F.B. (1987) Selection procedure for forest nature reserves in Nova Scotia, Canada. *Biological Conservation*, **41**, 185–201.

Goldsmith, F.B. (ed.) (1991a) *Monitoring for Conservation and Ecology*, Chapman & Hall, London.

Goldsmith, F.B. (1991b) The selection of protected areas, in *The Scientific Management of Temperate Communities for Conservation* (eds I.F. Spellerberg, F.B. Goldsmith and M.G. Morris), Blackwell Scientific, Oxford, pp. 273–91.

Good, R. (1948) *A Geographical Handbook of the Dorset Flora*, The Dorset Natural History and Archaeological Society, Dorchester.

Goodman, S.M., Meininger, P.L., Baha el din, S.H. *et al.* (1989) *The Birds of Egypt*, Oxford University Press, Oxford.

Gotelli, N.J. (1991) Metapopulation models: the rescue effect, the propagule rain, and the core-satellite hypothesis. *American Naturalist*, **138**, 768–76.

Gotelli, N.J. and Graves, G.R. (1990) Body size and the occurrence of avian species on land-bridge islands. *Journal of Biogeography*, **17**, 315–25.

Gotelli, N.J. and Simberloff, D. (1987) The distribution and abundance of tallgrass prairie plants: A test of the core-satellite hypothesis. *American Naturalist*, **130**, 18–35.

Graber, R.E. (1980) The life history and ecology of *Potentilla robbinsiana*. *Rhodora*, **82**, 131–40.

Graves, G.R. (1988) Linearity of geographic range and its possible effect on the population structure of Andean birds. *The Auk*, **105**, 47–52.

Gray, J.S. (1987) Species–abundance patterns, in *Organisation of Communities: Past and Present* (eds J.H.R. Gee and P.S. Giller), Blackwell Scientific, Oxford, pp. 53–67.

Gray, R.D. and Craig, J.L. (1991) Theory really matters: hidden assumptions in the concept of 'habitat requirements'. *Acta XX Congressus Internationalis Ornithologici*, 2553–60.

Green, R.E. and Hirons, G.J.M. (1991) The relevance of population studies to the conservation of threatened birds, in *Bird Population Studies: Relevance to Conservation and Management* (eds C.M. Perrins, J.-D. Lebreton and G.J.M. Hirons), Oxford University Press, Oxford, pp. 594–633.

Gregory, R.D., Keymer, A.E. and Harvey, P.H. (1991) Life history, ecology and parasite community structure in Soviet birds. *Biological Journal of the Linnean Society*, **43**, 249–62.

Greig-Smith, J. and Sagar, G.R. (1981) Biological causes of local rarity in *Carlina vulgaris*, in *The Biological Aspects of Rare Plant Conservation* (ed. H. Synge), Wiley, New York, pp. 389–400.

Griggs, R.F. (1940) The ecology of rare plants. *Bulletin of the Torrey Botanical Club*, **67**, 575–94.

Grinnell, J. (1922) The role of the 'accidental'. *The Auk*, **39**, 373–80.

Guerrant, E.O., Jr. (1992) Genetic and demographic considerations in the sampling and reintroduction of rare plants, in *Conservation Biology: The Theory and Practice of*

Nature Conservation, Preservation and Management (eds P.L. Fiedler and S.K. Jain), Chapman & Hall, London, pp. 321–44.

Haila, Y. (1988) Calculating and miscalculating density: the role of habitat geometry. *Ornis Scandinavica*, **19**, 88–92.

Haila, Y. and Järvinen, O. (1981) The underexploited potential of bird census in insular ecology. *Studies in Avian Biology*, **6**, 559–65.

Haila, Y. and Järvinen, O. (1983) Land bird communities on a Finnish island: species impoverishment and abundance patterns. *Oikos*, **41**, 255–73.

Haila, Y., Järvinen, O. and Koskimies, P. (eds) (1989) Monitoring bird populations in varying environments. *Annales Zoologici Fennici*, **26**, 149–330.

Hall, B.P. and Moreau, R.M. (1962) A study of rare birds of Africa. *Bulletin of the British Museum (Natural History) Zoology*, **8**, 313–78.

Hansen, R.M. and Ueckert, D.N. (1970) Dietary similarity of some primary consumers. *Ecology*, **51**, 640–8.

Hansen, T.A. (1978) Larval dispersal and species longevity in Lower Tertiary gastropods. *Science (Wash.)*, **199**, 885–7.

Hansen, T.A. (1980) Influence of larval dispersal and geographic distribution on species longevity in neogastropods. *Paleobiology*, **6**, 193–207.

Hanski, I. (1982a) Dynamics of regional distribution: the core and satellite species hypothesis. *Oikos*, **38**, 210–21.

Hanski, I. (1982b) Communities of bumblebees: testing the core-satellite species hypothesis. *Annales Zoologici Fennici*, **19**, 65–73.

Hanski, I. (1982c) Distributional ecology of anthropochorous plants in villages surrounded by forest. *Annales Botanici Fennici*, **19**, 1–15.

Hanski, I. (1991a) Single-species metapopulation dynamics: concepts, models and observations. *Biological Journal of the Linnean Society*, **42**, 17–38.

Hanski, I. (1991b) North temperate dung beetles, in *Dung Beetle Ecology* (eds I. Hanski and Y. Cambefort), Princeton University Press, Princeton, NJ, pp. 75–96.

Hanski, I. (1991c) Reply to Nee, Gregory and May. *Oikos*, **62**, 88–9.

Hanski, I. and Cambefort, Y. (1991) Spatial processes, in *Dung Beetle Ecology* (eds I. Hanski and Y. Cambefort), Princeton University Press, Princeton, NJ, pp. 283–304.

Hanski, I. and Koskela, H. (1978) Stability, abundance, and niche width in the beetle community inhabiting cow dung. *Oikos*, **31**, 290–8.

Hanski, I., Kouki, J. and Halkka, A. (1993) Three explanations of the positive relationship between distribution and abundance of species, in *Historical and Geographical Determinants of Community Diversity* (eds R. Ricklefs and D. Schluter), University of Chicago Press, Chicago, IL, in press.

Harding, P.T. (1991) National species distribution surveys, in *Monitoring for Conservation and Ecology* (ed. F.B. Goldsmith), Chapman & Hall, London, pp. 133–54.

Harmsen, R. (1983) Abundance distribution and evolution of community structure. *Evolutionary Theory*, **6**, 283–92.

Harper, J.L. (1981) The meanings of rarity, in *The Biological Aspects of Rare Plant Conservation* (ed. H. Synge), Wiley, New York, pp. 189–203.

Harper, K.T. (1979) Some reproductive and life history characteristics of rare plants and implications of management. *Great Basin Naturalist Memoirs*, **3**, 129–37.

Harrison, J.A. (1989) Atlassing as a tool in conservation, with special reference to the Southern African Bird Atlas Project, in *Biotic Diversity in Southern Africa: concepts and conservation* (ed. B.J. Huntley), Oxford University Press, Oxford, 157–69.

Harrison, S. (1991) Local extinction in a metapopulation context: an empirical evaluation. *Biological Journal of the Linnean Society*, **42**, 73–88.

Hartshorn, G.S. and Poveda, L.J. (1983) Checklist of trees, in *Costa Rican Natural History* (ed. D.H. Janzen), University of Chicago Press, Chicago, IL, pp. 158–83.

Harvey, P.H. and Pagel, M.D. (1991) *The Comparative Method in Evolutionary Biology*, Oxford University Press, Oxford.

Heath, J., Pollard, E. and Thomas, J.A. (1984) *Atlas of Butterflies in Britain and Ireland*, Penguin Books, Harmondsworth.

Hedderson, T.A. (1992) Rarity at range limits; dispersal capacity and habitat relationships of extraneous moss species in a boreal Canadian National Park. *Biological Conservation*, **59**, 113–120.

Hegazy, A.K. and Eesa, N.M. (1991) On the ecology, insect seed-predation, and conservation of a rare and endemic plant species: *Ebenus armitagei* (Leguminosae). *Conservation Biology*, **5**, 317–24.

Hengeveld, R. (1990) *Dynamic Biogeography*, Cambridge University Press, Cambridge.

Hengeveld, R. and Haeck, J. (1981) The distribution of abundance. II. Models and implications. *Proceedings of the Koninklijke Nederlandse Akademie van Wetenschappen*, **C84**, 257–84.

Hengeveld, R. and Haeck, J. (1982) The distribution of abundance. I. Measurements. *Journal of Biogeography*, **9**, 303–16.

Hengeveld, R. and Hogeweg, P. (1979) Cluster analysis of the distribution patterns of Dutch carabid species (Col.), in *Multivariate Methods in Ecological Work* (eds L. Orloci, C.R. Rao and W.M. Stiteler), International Co-operative Publishing House, Fairland, MD, pp. 65–86.

Hergstrom, K. and Niall, R. (1990) Presence–absence sampling of twospotted spider mite (Acari: Tetranychidae) in pear orchards. *Journal of Economic Entomology*, **83**, 2032–5.

Heywood, V.H. (1988) Rarity: a privilege and a threat, in *Proceedings of the XIV International Congress* (eds W. Greuter and B. Zimmer), Koeltz, Konigstein, Taunus, pp. 277–90.

Hill, M.O. (1991) Patterns of species distribution in Britain elucidated by canonical correspondence analysis. *Journal of Biogeography*, **18**, 247–55.

Hill, N.M. and Keddy, P.A. (1992) Prediction of rarities from habitat variables: coastal plain plants on Nova Scotian lakeshores. *Ecology*, **73**, 1852–9.

Hill, N.P. and Hagan, J.M. III (1991) Population trends of some northeastern North American landbirds: a half-century of data. *The Wilson Bulletin*, **103**, 165–82.

Hockin, D.C. (1981) The environmental determinants of the insular butterfly faunas of the British Isles. *Biological Journal of the Linnean Society*, **16**, 63–70.

Hodgson, J.G. (1986a) Commonness and rarity in plants with special reference to the Sheffield flora. Part I: The identity, distribution and habitat characteristics of the common and rare species. *Biological Conservation*, **36**, 199–252.

Hodgson, J.G. (1986b) Commonness and rarity in plants with special reference to the Sheffield flora. Part II: The relative importance of climate, soils and land use. *Biological Conservation*, **36**, 253–74.

Hodgson, J.G. (1986c) Commonness and rarity in plants with special reference to the Sheffield flora. Part III. Taxonomic and evolutionary aspects. *Biological Conservation*, **36**, 275–96.

Hodgson, J.G. (1986d) Commonness and rarity in plants with special reference to the

Sheffield flora. Part IV. A European context with particular reference to endemism. *Biological Conservation*, **36**, 297–314.

Hodgson, J.G. (1991) Management for the conservation of plants with particular reference to the British flora, in *The Scientific Management of Temperate Communities for Conservation* (eds I.F. Spellerberg, F.B. Goldsmith and M.G. Morris), Blackwell Scientific, Oxford, pp. 81–102.

Hodgson, J.G. and Grime, J.P. (1990) The role of dispersal mechanisms, regenerative strategies and seed banks in the vegetation dynamics of the British landscape, in *Species Dispersal in Agricultural Habitats* (eds R.G.H. Bunce and D.C. Howard), Belhaven Press, London, pp. 65–81.

Holmes, R.T. and Sherry, T.W. (1988) Assessing population trends of New Hampshire forest birds: local vs. regional patterns. *The Auk*, **105**, 756–68.

Holsinger, K.E. and Gottlieb, L.D. (1991) Conservation of rare and endangered plants: principles and prospects, in *Genetics and Conservation of Rare Plants* (eds D.A. Falk and K.E. Holsinger), Oxford University Press, Oxford, pp. 195–208.

Howe, R.W., Davis, G.J. and Mosca, V. (1991) The demographic significance of 'sink' populations. *Biological Conservation*, **57**, 239–55.

Hubbell, S.P. and Foster, R.B. (1986) Commonness and rarity in a Neotropical forest: implications for tropical tree conservation, in *Conservation Biology: the Science of Scarcity and Diversity* (ed. M.A. Soulé), Sinauer Associates, Massachusetts, pp. 205–231.

Hughes, R.G. (1986) Theories and models of species abundance. *American Naturalist*, **128**, 879–99.

Hunter, M.L., Jr., Jacobson, G.L. Jr., and Webb, T. III (1988) Paleoecology and the coarse-filter approach to maintaining biological diversity. *Conservation Biology*, **2**, 375–85.

Huntley, B. (1988) Glacial and Holocene vegetation history: Europe, in *Vegetation History* (eds B. Huntley and T. Webb III), Kluwer, Dordrecht, pp. 341–83.

Huntley, B. (1991) Historical lessons for the future, in *The Scientific Management of Temperate Communities for Conservation* (eds I.F. Spellerberg, F.B. Goldsmith and M.G. Morris), Blackwell Scientific, Oxford, pp. 473–503.

Huntley, B. and Birks, H.J.B. (1983) *An Atlas of Past and Present Pollen Maps for Europe: 0–13 000 Years Ago*, Cambridge University Press, Cambridge.

Huntley, B. and Webb, T. III (1989) Migration: species' response to climatic variations caused by changes in the earth's orbit. *Journal of Biogeography*, **16**, 5–19.

Hutchings, M.J. (1987a) The population biology of the early spider orchid, *Ophrys sphegodes* Mill. I. A demographic study from 1975 to 1984. *Journal of Ecology*, **75**, 711–27.

Hutchings, M.J. (1987b) The population biology of the early spider orchid, *Ophrys sphegodes* Mill. II. Temporal patterns in behaviour. *Journal of Ecology*, **75**, 729–42.

Itamies, J. (1983) Factors contributing to the succession of plants and Lepidoptera on the islands off Rauma, SW Finland. *Acta Universitatis Ouluensis*, **18**, 1–48.

Jablonski, D. (1986a) Causes and consequences of mass extinctions: a comparative approach, in *Dynamics of Extinction* (ed. D.K. Elliot), Wiley, New York, pp. 183–229.

Jablonski, D. (1986b) Background and mass extinction: the alternation of macroevolutionary regimes. *Science (Wash.)*, **231**, 129–33.

Jablonski, D. (1987) Heritability at the species level: Analysis of geographic ranges of Cretaceous mollusks. *Science (Wash.)*, **238**, 360–63.

Jablonski, D. (1988) Response [to Russell and Lindberg]. *Science (Wash.)*, **240**, 969.

Jablonski, D. (1989) The biology of mass extinction: a palaeontological view. *Philosophical Transactions of the Royal Society of London (Series B)*, **325**, 357–68.

Jablonski, D. and Valentine, J.W. (1990) From regional to total geographic ranges: testing the relationship in Recent bivalves. *Paleobiology*, **16**, 126–42.

Jackson, J.B.C. (1974) Biogeographic consequences of eurytopy and stenotopy among marine bivalves and their evolutionary significance. *American Naturalist*, **108**, 541–60.

Jarman, P.J. (1974) The social organisation of antelopes in relation to their ecology. *Behaviour*, **48**, 215–67.

Järvinen, O. (1982) Conservation of endangered plant populations: single large or several small reserves. *Oikos*, **38**, 301–7.

Jefferson, R.G. and Usher, M.B. (1986) Ecological succession and the evaluation of non-climax communities, in *Wildlife Conservation Evaluation* (ed. M.B. Usher), Chapman & Hall, London, pp. 69–91.

Jenkins, R.A., Wade, K.R. and Pugh, E. (1984) Macroinvertebrate–habitat relationships in the River Teifi catchment and the significance to conservation. *Freshwater Biology*, **14**, 23–42.

Juliano, S.A. (1983) Body size, dispersal ability, and range size in North American species of *Brachinus* (Coleoptera: Carabidae). *Coleopterists Bulletin*, **37**, 232–8.

Karr, J.R. (1977) Ecological correlates of rarity in a tropical forest bird community. *The Auk*, **94**, 240–7.

Karr, J.R. (1982a) Population variability and extinction in the avifauna of a tropical land bridge island. *Ecology*, **63**, 1975–8.

Karr, J.R. (1982b) Avian extinction on Barro Colorado Island, Panama: A reassessment. *American Naturalist*, **119**, 220–39.

Karron, J.D. (1987) The pollination ecology of co-occurring geographically restricted and widespread species of *Astragalus* (Fabaceae). *Biological Conservation*, **39**, 179–93.

Kattan, G.H. (1992) The birds of the Cordillera Central of Colombia. *Conservation Biology*, **6**, 64–70.

Kavanaugh, D.H. (1985) On wing atrophy in carabid beetles (Coleoptera: Carabidae), with special reference to Nearctic *Nebria*, in *Taxonomy, Phylogeny and Zoogeography of Beetles and Ants* (ed. G.E. Ball), Dr. W. Junk Publishers, Dordrecht, pp. 408–31.

Keddy, P.A. (1989) *Competition*, Chapman & Hall, London.

Kelcey, J.G. (1984) Industrial development and the conservation of vascular plants, with special reference to Britain. *Environmental Conservation*, **11**, 235–45.

Kemp, W.P., Harvey, S.J. and O'Neill, K.M. (1990) Patterns of vegetation and grasshopper community composition. *Oecologia (Berl.)*, **83**, 299–308.

Kennedy, C.E.J. and Southwood, T.R.E. (1984) The number of species of insects associated with British trees: a re-analysis. *Journal of Animal Ecology*, **53**, 455–78.

Kershaw, K.A. and Looney, J.H.H. (1985) *Quantitative and Dynamic Plant Ecology*, Arnold, London.

Kirkpatrick, J.B. (1983) An iterative method for establishing priorities for the selection of nature reserves: an example from Tasmania. *Biological Conservation*, **25**, 127–34.

Koch, C.F. (1987) Prediction of sample size effects on the measured temporal and geographic distribution patterns of species. *Paleobiology*, **13**, 100–7.

Kolasa, J. and Strayer, D. (1988) Patterns of the abundance of species: a comparison of two hierarchical models. *Oikos*, **53**, 235–41.

Kouki, J. and Hüyrinen, U. (1991) On the relationship between distribution and abundance in birds breeding on Finnish mires: the effect of habitat specialization. *Ornis Fennica*, **68**, 170–7.

Krebs, C.J. (1978) *Ecology: The Experimental Analysis of Distribution and Abundance*, Harper & Row, New York.

Krebs, C.J. (1989) *Ecological Methodology*, Harper & Row, New York.

Kruckeberg, A.R. and Rabinowitz, D. (1985) Biological aspects of endemism in higher plants. *Annual Review of Ecology and Systematics*, **16**, 447–79.

Kubitzki, K. (1977) The problem of rare and of frequent species: the monographer's view, in *Extinction is Forever* (eds G.T. Prance and T.S. Elias), The New York Botanical Garden, New York, pp. 331–6.

Kunin, W.E. and Gaston, K.J. (1993) The biology of rarity: patterns, causes, and consequences. *Trends in Ecology and Evolution*, **8**, 298–301.

Kuno, E. (1986) Evaluation of statistical precision and design of efficient sampling for the population estimation based on frequency of occurrence. *Researches on Population Ecology*, **28**, 305–19.

Lacy, R.C. and Bock, C.E. (1986) The correlation between range size and local abundance of some North American birds. *Ecology*, **67**, 258–60.

Lahti, T., Kemppainen, E., Kurtto, A. and Uotila, P. (1991) Distribution and biological characteristics of threatened vascular plants in Finland. *Biological Conservation*, **55**, 299–314.

Lamb, C. (1810) (Letter to T. Manning, 2 Jan., 1810; cited in Cohen, J.M. and Cohen, M.J. (1979). *The Penguin Dictionary of Quotations*, Penguin Books, Harmondsworth.)

Landa, K. and Rabinowitz, D. (1983) Relative preference of *Arphia sulphurea* (Orthoptera: Acrididae) for sparse and common prairie grasses. *Ecology*, **64**, 392–5.

Lande, R. (1988) Genetics and demography in biological conservation. *Science (Wash.)*, **241**, 1455–60.

Landolt, E. (1991) Distribution patterns of flowering plants in the city of Zurich, in *Modern Ecology: Basic and Applied Aspects* (eds G. Esser and D. Overdieck), Elsevier, Amsterdam, pp. 807–22.

Laurance, W.F. (1991) Ecological correlates of extinction proneness in Australian tropical rain forest mammals. *Conservation Biology*, **5**, 79–89.

Lawton, J.H. (1989) What is the relationship between population density and body size in animals? *Oikos*, **55**, 429–34.

Lawton, J.H. (1990) Species richness and population dynamics of animal assemblages. Patterns in body size, abundance, space. *Philosophical Transactions of the Royal Society of London (Series B)*, **330**, 283–91.

Lawton, J.H. (1991) Are species useful? *Oikos*, **62**, 3–4.

Lawton, J.H. and Gaston, K.J. (1989) Temporal patterns in the herbivorous insects of bracken: a test of community predictability. *Journal of Animal Ecology*, **58**, 1021–34.

Lawton, J.H. and Schröder, D. (1977) Effects of plant type, size of geographical range and taxonomic isolation on number of insect species associated with British plants. *Nature*, **265**, 137–40.

Lawton, J.H. and Woodroffe, G.L. (1991) Habitat and the distribution of water voles: why are there gaps in a species' range? *Journal of Animal Ecology*, **60**, 79–91.

Leather, S.R. (1985) Does the bird cherry have its 'fair share' of insect pests? An appraisal of the species–area relationships of the phytophagous insects associated with British *Prunus* species. *Ecological Entomology*, **10**, 43–56.

Leck, C.F. (1979) Avian extinctions in an isolated tropical wet-forest preserve, Ecuador. *The Auk*, **96**, 343–52.

Legendre, L. and Legendre, P. (1983) *Numerical Ecology*, Elsevier, Amsterdam.

Leigh, E.G. (1981) The average life time of a population in a varying environment. *Journal of Theoretical Biology*, **90**, 213–39.

Lesica, P. (1992) Autecology of the endangered plant *Howellia aquatilis*; implications for management and reserve design. *Ecological Applications*, **2**, 411–21.

Levin, S.A. (1992) The problem of pattern and scale in ecology. *Ecology*, **73**, 1943–67.

Levins, R. (1968) *Evolution in Changing Environments*, Princeton University Press, Princeton, NJ.

Levins, R. (1969) Some demographic and genetic consequences of environmental heterogeneity for biological control. *Bulletin of the Entomological Society of America*, **15**, 237–40.

Lindenmayer, D.B., Nix, H.A., McMahon, J.P. *et al.* (1991) The conservation of Leadbeater's possum, *Gymnobelideus leadbeateri* (McCoy): a case study of the use of bioclimatic modelling. *Journal of Biogeography*, **18**, 371–83.

Lindstedt, S.L. and Boyce, M.S. (1985) Seasonality, fasting endurance, and body size in mammals. *American Naturalist*, **125**, 873–8.

Lloyd, C., Tasker, M.L. and Partridge, K. (1991) *The Status of Seabirds in Britain and Ireland*, Poyser, London.

Longmore, R. (ed.) (1986) *Snakes: Atlas of Elapid Snakes of Australia*, Australian Government Publishing Service, Canberra.

Longton, R.E. (1992) Reproduction and rarity in British mosses. *Biological Conservation*, **59**, 89–98.

Lovejoy, T.E. (1988) Diverse considerations, in *Biodiversity* (ed. E.O. Wilson), National Academy Press, Washington, pp. 421–7.

Lubchenco, J. (1978) Plant species diversity in a marine intertidal community: importance of herbivore food preference and algal competitive abilities. *American Naturalist*, **112**, 23–39.

Ludwig, J.A. and Reynolds, J.F. (1988) *Statistical Ecology*, Wiley, New York.

Lumaret, J. and Kirk, A.A. (1991) South temperate dung beetles, in *Dung Beetle Ecology* (eds I. Hanski and Y. Cambefort), Princeton University Press, Princeton, NJ, pp. 97–115.

MacArthur, R.H. (1972) *Geographical Ecology*, Harper & Row, New York.

Mace, G.M. and Lande, R. (1991) Assessing extinction threats: toward a re-evaluation of IUCN threatened species categories. *Conservation Biology*, **5**, 148–57.

Macior, L.W. (1978) The pollination ecology and endemic adaptation of *Pedicularis furbishae* S. Wats. *Bulletin of the Torrey Botanical Club*, **105**, 268–77.

Macior, L.W. (1980) Population ecology of the furbish lousewort, *Pedicularis furbishae* S. Wats. *Rhodora*, **82**, 105–11.

MacKinnon, J. and MacKinnon, K. (1986a) *Review of the Protected Areas System in the Indo-malayan Realm*, IUCN, Gland.

MacKinnon, J. and MacKinnon, K. (1986b) *Review of the Protected Areas System in the Afrotropical Realm*, IUCN and UNEP, Cambridge.

Mac Nally, R.C. (1989) The relationship between habitat breadth, habitat position, and abundance in forest and woodland birds along a continental gradient. *Oikos*, **54**, 44–54.

Mac Nally, R.C. and Doolan, J.M. (1986) An empirical approach to guild structure: habitat relationships in nine species of eastern-Australian cicadas. *Oikos*, **47**, 33–46.

Macpherson, E. (1989) Influence of geographical distribution, body size and diet on population density of benthic fishes off Namibia (South West Africa). *Marine Ecology Progress Series*, **50**, 295–9.

Magurran, A.E. (1988) *Ecological Diversity and its Measurement*, Croom Helm, London.

Main, A. (1982) Rare species: precious or dross? in *Species at Risk: Research in Australia* (eds R.H. Groves and W.D.L. Ride), Springer-Verlag, New York, pp. 163–74.

Main, A.R. (1984) Rare species: problems of conservation. *Search*, **15**, 93–7.

Maitland, P.S. (1985) Criteria for the selection of important sites for freshwater fish in the British Isles. *Biological Conservation*, **31**, 335–53.

Mandelbrot, B.B. (1967) How long is the coast of Britain? Statistical self-similarity and fractional dimension. *Science (Wash.)*, **155**, 636–8.

Mandelbrot, B.B. (1982) *The Fractal Geometry of Nature*, W.H. Freeman, San Francisco.

Marchant, J.H., Hudson, R., Carter, S.P. and Whittington, P. (1990) *Population Trends in British Breeding Birds*, British Trust for Ornithology, Tring, Hertfordshire.

Margules, C. and Usher, M.B. (1981) Criteria used in assessing wildlife conservation potential: a review. *Biological Conservation*, **21**, 79–109.

Margules, C.R., Nicholls, A.O. and Pressey, R.L. (1988) Selecting networks of reserves to maximise biological diversity. *Biological Conservation*, **43**, 63–76.

Margules, C.R. and Stein, J.L. (1989) Patterns in the distribution of species and the selection of nature reserves: An example from *Eucalyptus* forests in south-eastern New South Wales. *Biological Conservation*, **50**, 219–38.

Marquet, P.A., Navarrete, S.A. and Castilla, J.C. (1990) Scaling population density to body size in rocky intertidal communities. *Science (Wash.)*, **250**, 1125–27.

Marquiss, M. (1989) Grey herons *Ardea cinerea* breeding in Scotland: numbers, distribution, and census techniques. *Bird Study*, **36**, 181–91.

Mason, C.F. (1990) Assessing population trends of scarce birds using information in a county bird report and archive. *Biological Conservation*, **52**, 303–20.

Mason, H.L. (1946a) The edaphic factor in narrow endemism. I. The nature of environmental influences. *Madrono*, **8**, 209–26.

Mason, H.L. (1946b) The edaphic feature in narrow endemism. II. The geographic occurrence of plants of highly restricted patterns of distribution. *Madrono*, **8**, 241–72.

Maurer, B.A. (1991) Concluding remarks: birds, body size, and evolution. *Acta XX Congressus Internationalis Ornithologici*, 835–7.

Maurer, B.A. and Brown, J.H. (1988) Distribution of energy use and biomass among species of North American terrestrial birds. *Ecology*, **69**, 1923–32.

Maurer, B.A., Ford, H.A. and Rapoport, E.H. (1991) Extinction rate, body size, and avifaunal diversity. *Acta XX Congressus Internationalis Ornithologici*, 826–34.

May, R.M. (1975) Patterns of species abundance and diversity, in *Ecology and Evolution of Communities* (eds M.L. Cody and J.M. Diamond), Harvard University Press, Cambridge, MA, pp. 81–120.

Mayfield, H.F. (1983) Kirtland's warbler, victim of its own rarity? *The Auk*, **100**, 974–6.

Mayr, E. (1963) *Animal Species and Evolution*, The Belknap Press of Harvard University Press, Cambridge, MA.

McAllister, D.E., Platania, S.P., Schueler, F.W. *et al.* (1986) Ichthyofaunal patterns on a geographical grid, in *Zoogeography of Freshwater Fishes of North America* (eds C.H. Hocutt and E.D. Wiley), Wiley, New York, pp. 17–51.

McArdle, B.H. (1990) When are rare species not there? *Oikos*, **57**, 276–7.

McArdle, B.H. and Gaston, K.J. (1992) Comparing population variabilities. *Oikos*, **64**, 610–12.

McArdle, B.H. and Gaston, K.J. (1993) The temporal variability of populations. *Oikos*, **67**, 187–91.

McArdle, B.H., Gaston, K.J. and Lawton, J.H. (1990) Variation in the size of animal populations: patterns, problems and artefacts. *Journal of Animal Ecology*, **59**, 439–54.

McClure, M.S. and Price, P.W. (1976) Ecotope characteristics of coexisting *Erythroneura* leafhoppers (Homoptera: Cicadellidae) on sycamore. *Ecology*, **57**, 928–40.

McCoy, E.D. (1990) The distribution of insects along elevational gradients. *Oikos*, **58**, 313–22.

McCoy, E.D. and Connor, E.F. (1980) Latitudinal gradients in the species diversity of North American mammals. *Evolution*, **34**, 193–203.

McCoy, E.D. and Mushinsky, H.R. (1992) Rarity of organisms in the sand pine scrub habitat of Florida. *Conservation Biology*, **6**, 537–48.

McGowan, J.A. and Walker, P.W. (1979) Structure in the copepod community of the North Pacific Central Gyre. *Ecological Monographs*, **49**, 195–226.

McIntosh, R.P. (1962) Raunkiaer's 'law' of frequency. *Ecology*, **43**, 533–5.

McIntyre, S. (1992) Risks associated with the setting of conservation priorities from rare plant species lists. *Biological Conservation*, **60**, 31–7.

McLaughlin, S.P. (1992) Are floristic areas hierarchically arranged? *Journal of Biogeography*, **19**, 21–32.

McNaughton, S.J. and Wolf, L.L. (1970) Dominance and the niche in ecological systems. *Science (Wash.)*, **167**, 131–9.

Meagher, T.R., Antonovics, J. and Primack, R. (1978) Experimental ecological genetics in *Plantago*, III. Genetic variation and demography in relation to survival of *Plantago cordata*, a rare species. *Biological Conservation*, **14**, 243–57.

Meffe, G.K. and Sheldon, A.L. (1990) Post-defaunation recovery of fish assemblages in southeastern blackwater streams. *Ecology*, **71**, 657–67.

Merikallio, E. (1951) On the numbers of land-birds in Finland. *Acta Zoologica Fennica*, **65**, 1–16.

Merikallio, E. (1958) Finnish birds, their distribution and numbers. *Fauna Fennica*, **5**, 1–181.

Miller, R.I. (1986) Predicting rare plant distribution patterns in the southern Appalachians of the southeastern USA. *Journal of Biogeography*, **13**, 293–311.

Miller, R.I., Bratton, S.P. and White, P.S. (1987) A regional strategy for reserve design and placement based on an analysis of rare and endangered species' distribution patterns. *Biological Conservation*, **39**, 255–68.

Miller, R.I., Stuart, S.N. and Howell, K.M. (1989) A methodology for analyzing rare species distribution patterns utilizing GIS technology: The rare birds of Tanzania. *Landscape Ecology*, **2**, 173–89.

Mitchley, J. and Grubb, P.J. (1986) Control of relative abundance of perennials in chalk grassland in southern England. I. Constancy of rank order and results of pot- and field-experiments on the role of interference. *Journal of Ecology*, **74**, 1139–66.

Moll, E.J. and Gubb, A.A. (1981) Aspects of the ecology of *Staavia dodii* in the South Western Cape of South Africa, in *The Biological Aspects of Rare Plant Conservation* (ed. H. Synge), Wiley, New York, pp. 331–42.

Monro, J. (1967) The exploitation and conservation of resources by populations of insects. *Journal of Animal Ecology*, **36**, 531–47.

Moore, N.W. (1991) Observe extinction or conserve diversity? in *The Conservation of Insects and their Habitats* (eds N.M. Collins and J.A. Thomas), Academic Press, London, pp. 237–62.

Morgan, K.H., Vermeer, K. and McKelvey, R.W. (1991) *Atlas of Pelagic Birds of Western Canada*, Occasional Paper No. 72, Canadian Wildlife Service.

Morris, M.G. (1991) The management of reserves and protected areas, in *The Scientific Management of Temperate Communities for Conservation* (eds I.F. Spellerberg, F.B. Goldsmith and M.G. Morris), Blackwell Scientific, Oxford, pp. 323–47.

Morrision, M.L., Marcot, B.G. and Mannan, R.W. (1992) *Wildlife–Habitat Relationships: Concepts and Applications*, The University of Wisconsin Press, Madison, WI.

Morse, D.R., Stork, N.E. and Lawton, J.H. (1988) Species number, species abundance and body length relationships of arboreal beetles in Bornean lowland rain forest trees. *Ecological Entomology*, **13**, 25–37.

Moulton, M.P. and Pimm, S.L. (1986) Species introductions to Hawaii, in *The Ecology of Biological Invasions of North America and Hawaii* (eds H.A. Mooney and J.A. Drake), Springer-Verlag, New York, pp. 231–49.

Müllenberg, M., Leipold, D., Mader, H.J. and Steinhauer, B. (1977) Island ecology of arthropods. II. Niches and relative abundances of Seychelles ants (Formicidae) in different habitats. *Oecologia (Berl.)*, **29**, 135–44.

Munton, P. (1987) Concepts of threat to the survival of species used in Red Data books and similar compilations, in *The Road to Extinction* (eds R. Fitter and M. Fitter), IUCN/UNEP, Gland, pp. 72–95.

Munves, J. (1975) Birds of a highland clearing in Cundinamacra, Colombia. *The Auk*, **92**, 307–321.

Murray, B.G. (1987) On the meaning of density dependence. *Oecologia (Berl.)*, **53**, 370–3.

Murray, M.D. and Nix, H.A. (1987) Southern limits of distribution and abundance of the biting-midge *Culicoides brevitarsis* Kieffer (Diptera: Ceratopogonidae) in southeastern Australia: an application of the GROWEST model. *Australian Journal of Zoology*, **35**, 575–85.

Nachman, G. (1981) A mathematical model of the functional relationship between density and spatial distribution of a population. *Journal of Animal Ecology*, **50**, 453–60.

Nachman, G. (1984) Estimates of mean population density and spatial distribution of *Tetranychus urticae* (Acarina: Tetranychidae) and *Phytoseiulus persimilis* (Acarina: Phytoseiidae) based upon the proportion of empty sampling units. *Journal of Applied Ecology*, **21**, 903–13.

National Aeronautics and Space Administration (1988) *Earth System Science*, NASA, Washington, DC.

Nee, S., Gregory, R.D. and May, R.M. (1991a) Core and satellite species: theory and artefacts. *Oikos*, **62**, 83–7.

Nee, S., Harvey, P.H. and May, R.M. (1991b) Lifting the veil on abundance patterns. *Proceedings of the Royal Society of London B*, **243**, 161–3.

Nee, S., Read, A.F., Greenwood, J.J.D. and Harvey, P.H. (1991c) The relationship between abundance and body size in British birds. *Nature*, **351**, 312–13.

Neuvonen, S. and Niemelä, P. (1981) Species richness of macrolepidoptera on Finnish deciduous trees and shrubs. *Oecologia (Berl.)*, **51**, 364–70.

182   References

Nicholls, A.O. (1989) How to make biological surveys go further with generalised linear models. *Biological Conservation*, **50**, 51–75.

Niemelä, P. and Neuvonen, S. (1983) Species richness of herbivores on hosts: how robust are patterns revealed by analysing published host plant lists? *Annales Entomologici Fennici*, **49**, 95–9.

Nilsson, C. (1986) Methods of selecting lake shorelines as nature reserves. *Biological Conservation*, **35**, 269–91.

Nilsson, C., Grelsson, G., Johnasson, M. and Sperens, U. (1988) Can rarity and diversity be predicted in vegetation along river banks? *Biological Conservation*, **44**, 201–12.

Nilsson, S.G., Bengtsson, J. and Ås, S. (1988) Habitat diversity or area *per se*? Species richness of woodland plants, carabid beetles and land snails on islands. *Journal of Animal Ecology*, **57**, 685–704.

Novotný , V. (1991) Effect of habitat persistence on the relationship between geographic distribution and local abundance. *Oikos*, **61**, 431–3.

Oakwood, M., Jurado, E., Leishman, M. and Westoby, M. (1993) Geographic ranges of plant species in relation to dispersal morphology, growth form, and diaspore weight. *Journal of Biogeography*, pp. 20, 563–572.

Obeso, J.R. (1992) Geographic distribution and community structure of bumblebees in the northern Iberian peninsula. *Oecologia (Berl.)*, **89**, 244–52.

O'Neill, J.P. and Pearson, D.L. (1974) Estudio preliminar de las aves de Yarinacocha, Departamento de Loreto, Peru. *Publicaciones del Museo de Historia Natural 'Javier Prado' Zoologica Serie A*, **25**, 1–13.

Osborne, P.E. and Tigar, B.J. (1992a) Priorities for bird conservation in Lesotho, southern Africa. *Biological Conservation*, **61**, 159–69.

Osborne, P.E. and Tigar, B.J. (1992b) Interpreting bird atlas data using logistic models: an example from Lesotho, Southern Africa. *Journal of Applied Ecology*, **29**, 55–62.

Otte, D. (1976) Species richness patterns of New World desert grasshoppers in relation to plant diversity. *Journal of Biogeography*, **3**, 197–209.

Owen, J. and Gilbert, F.S. (1989) On the abundance of hoverflies (Syrphidae). *Oikos*, **55**, 183–93.

Owen-Smith, N. (1989) Megafaunal extinctions: the conservation message from 11,000 years B.P. *Conservation Biology*, **3**, 405–12.

Pagel, M.P., May, R.M. and Collie, A.R. (1991) Ecological aspects of the geographic distribution and diversity of mammal species. *American Naturalist*, **137**, 791–815.

Paine, M.D. (1990) Life history of darters (Percidae: Etheostomatiini) and their relationship with body size, reproductive behaviour, latitude and rarity. *Journal of Fish Biology*, **37**, 473–88.

Parrish, J.A.D. and Bazzaz, F.A. (1976) Underground niche separation in successional plants. *Ecology*, **57**, 1281–8.

Pearson, D.L. (1977) A pantropical comparison of bird community structure on six lowland forest sites. *Condor*, **79**, 232–44.

Perring, F.H. and Farrell, L. (1983) *British Red Data Books: 1. Vascular Plants* (2nd edn), RSNC, Lincoln.

Perring, F.H. and Walters, S.M. (eds) (1962) *Atlas of the British Flora*, Nelson, London.

Perry, J.N. (1987) Host-parasitoid models of intermediate complexity. *American Naturalist*, **130**, 955–7.

Perry, J.N. and Taylor, L.R. (1985) Adès: New ecological families of species-specific frequency distributions that describe repeated spatial samples with an intrinsic power-law variance-mean property. *Journal of Animal Ecology*, **54**, 931–53.

Perry, J.N. and Taylor, L.R. (1986) Stability of real interacting populations in space and time: implications, alternatives and the negative binomial k_c. *Journal of Animal Ecology*, **55**, 1053–68.

Peters, R.H. (1983) *The Ecological Implications of Body Size*, Cambridge University Press, Cambridge.

Peters, R.H. and Raelson, J.V. (1984) Relations between individual size and mammalian population density. *American Naturalist*, **124**, 498–517.

Peters, R.H. and Wassenberg, K. (1983) The effect of body size on animal abundance. *Oecologia (Berl.)*, **60**, 89–96.

Pianka, E.R. (1986) *Ecology and Natural History of Desert Lizards*, Princeton University Press, Princeton, NJ.

Pianka, E.R. and Schall, J.J. (1981) Species densities of Australian vertebrates, in *Ecological Biogeography of Australia* (ed. A. Keast), Dr W. Junk, The Hague, pp. 1675–94.

Pickard, J. (1983) Rare or threatened vascular plants of Lord Howe Island. *Biological Conservation*, **27**, 125–39.

Pielou, E.C. (1977) The latitudinal spans of seaweed species and their patterns of overlap. *Journal of Biogeography* **4**, 299–311.

Pielou, E.C. (1978) Latitudinal overlap of seaweed species: evidence for quasi-sympatric speciation. *Journal of Biogeography*, **5**, 227–38.

Pielou, E.C. (1979) *Biogeography*, Wiley, New York.

Pigott, C.D. (1981) The status, ecology and conservation of *Tilia platyphyllos* in Britain, in *The Biological Aspects of Rare Plant Conservation* (ed. H. Synge), Wiley, New York, pp. 305–17.

Pimentel, D. and Wheeler, A.G. (1973) Species and diversity of arthropods in the Alfalfa community. *Environmental Entomology*, **2**, 659–68.

Pimm, S.L. (1984) The complexity and stability of ecosystems. *Nature*, **307**, 321–6.

Pimm, S.L., Jones, H.L. and Diamond, J. (1988) On the risk of extinction. *American Naturalist*, **132**, 757–85.

Plant, C.W. (1987) *The Butterflies of the London Area*, London Natural History Society, London.

Pollard, E. (1979) Population ecology and change in range of the white admiral butterfly *Ladoga camilla* L. in England. *Ecological Entomology*, **4**, 61–74.

Pomeroy, D. and Ssekabiira, D. (1990) An analysis of the distributions of terrestrial birds in Africa. *African Journal of Ecology*, **28**, 1–13.

Pressey, R.L. and Nicholls, A.O. (1989a) Efficiency in conservation evaluation: scoring versus iterative approaches. *Biological Conservation*, **50**, 199–218.

Pressey, R.L. and Nicholls, A.O. (1989b) Application of a numerical algorithm to the collection of reserves in semi-arid New South Wales. *Biological Conservation*, **50**, 263–78.

Preston, F.W. (1948) The commonness and rarity of species. *Ecology*, **29**, 254–83.

Preston, F.W. (1958) Analysis of the Audubon Christmas counts in terms of the lognormal curve. *Ecology*, **39**, 620–4.

Preston, F.W. (1962) The canonical distribution of commonness and rarity. *Ecology*, **43**, 185–215, 410–32.

Preston, F.W. (1980) Noncanonical distributions of commonness and rarity. *Ecology*, **6**, 88–97.

Price, D. de Solla (1963) *Little Science, Big Science*, Columbia, New York.

Price, M.V. and Endo, P.R. (1989) Estimating the distribution and abundance of a cryptic species, *Dipodomys stephensi* (Rodentia: Heteromyidae), and implications for management. *Conservation Biology*, **3**, 293–301.

Price, P.W. (1971) Niche breadth and dominance of parasitic insects sharing the same host species. *Ecology*, **52**, 587–96.

Primack, R.B. (1980) Phenotypic variation of rare and widespread species of *Plantago*. *Rhodora*, **82**, 87–96.

Primack, R.B. and Miao, S.L. (1992) Dispersal can limit local plant distribution. *Conservation Biology*, **6**, 513–19.

Prober, S.M. (1992) Environmental influences on the distribution of the rare *Eucalyptus paliformis* and the common *E. fraxinoides*. *Australian Journal of Ecology*, **17**, 51–65.

Prober, S.M. and Austin, M.P. (1990) Habitat peculiarity as a cause of rarity in *Eucalyptus paliformis*. *Australian Journal of Ecology*, **16**, 189–205.

Pulliam, H.R. (1988) Sources, sinks, and population regulation. *American Naturalist*, **132**, 652–61.

Quinn, R.M., Lawton, J.H., Eversham, B.C. and Wood, S.N. (1994) The biogeography of scarce vascular plants in Britain with respect to habitat preference, dispersal ability and reproductive biology. *Biological Conservation*, in press.

Rabinowitz, D. (1978) Abundance and diaspore weight in rare and common prairie grasses. *Oecologia (Berl.)*, **37**, 213–19.

Rabinowitz, D. (1981a) Seven forms of rarity, in *The Biological Aspects of Rare Plant Conservation* (ed. H. Synge), Wiley, New York, pp. 205–17.

Rabinowitz, D. (1981b) Buried viable seeds in a North American tall-grass prairie: the resemblance of their abundance and composition to dispersing seeds. *Oikos*, **36**, 191–5.

Rabinowitz, D., Bassett, B.K. and Renfro, G.E. (1979) Abundance and neighbourhood structure for sparse and common grasses in a Missouri prairie. *American Journal of Botany*, **66**, 867–9.

Rabinowitz, D., Cairns, S. and Dillon, T. (1986) Seven forms of rarity and their frequency in the flora of the British Isles, in *Conservation Biology: the Science of Scarcity and Diversity* (ed. M.E. Soulé), Sinauer Associates, Sunderland, MA, pp. 182–204.

Rabinowitz, D. and Rapp, J.K. (1981) Dispersal abilities of seven sparse and common grasses from a Missouri prairie. *American Journal of Botany*, **68**, 616–24.

Rabinowitz, D. and Rapp, J.K. (1985) Colonization and establishment of Missouri prairie plants on artificial soil disturbances. III. Species abundance distributions, survivorship, and rarity. *American Journal of Botany*, **72**, 1635–40.

Rabinowitz, D., Rapp, J.K., Cairns, S. and Mayer, M. (1989) The persistence of rare prairie grasses in Missouri: environmental variation buffered by reproductive output of sparse species. *American Naturalist*, **134**, 525–44.

Rabinowitz, D., Rapp, J.K. and Dixon, P.M. (1984) Competitive abilities of sparse grass species: means of persistence or cause of abundance. *Ecology*, **65**, 1144–54.

Rahel, F.J. (1990) The hierarchical nature of community persistence: a problem of scale. *American Naturalist*, **136**, 328–44.

Ralph, C.J. and Scott, J.M. (eds) (1981) Estimating numbers of terrestrial birds. *Studies in Avian Biology*, **6**.

Rands, A.S. and Myers, C.W. (1990) The herpetofauna of Barro Colorado Island, Panama: an ecological summary, in *Four Neotropical Forests* (ed. A.H. Gentry), Yale University Press, New Haven, CT, pp. 386–409.

Rapoport, E.H. (1982) *Areography: Geographical Strategies of Species*, Pergamon, Oxford.

Rapoport, E.H., Borioli, G., Monjeau, J.A. *et al.* (1986) The design of nature reserves: a simulation trial for assessing specific conservation value. *Biological Conservation*, **37**, 269–90.

Raunkiaer, C. (1934) *The Life Forms of Plants and Statistical Plant Geography*, Clarendon, Oxford.

Reaka, M.L. (1980) Geographic range, life history patterns, and body size in a guild of coral-dwelling mantis shrimps. *Evolution*, **34**, 1019–30.

Reed, J.M. (1992) A system for ranking conservation priorities for Neotropical migrant birds based on relative susceptibility to extinction, in *Ecology and Conservation of Neotropical Migrant Landbirds* (eds J.M. Hagan III and D.W. Johnston), Smithsonian Institution Press, Washington, pp. 524–36.

Reed, T.M. (1982) The number of butterfly species on British islands. *Proceedings of the 3rd Congress of European Lepidopterists*, Cambridge, pp. 146–52.

Reveal, J.L. (1981) The concepts of rarity and population threats in plant communities, in *Rare Plant Conservation* (eds L.E. Morse and M.S. Henefin), The New York Botanical Garden, Bronx, pp. 41–6.

Ricklefs, R.E. (1972) Dominance and the niche in bird communities. *American Naturalist*, **106**, 538–45.

Ricklefs, R.E. and Latham, R.E. (1992) Intercontinental correlation of geographic ranges suggests stasis in ecological traits of relict genera of temperate perennial herbs. *American Naturalist*, **139**, 1305–21.

Ridgely, R.S. (1976) *A Guide to the Birds of Panama*, Princeton University Press, Princeton, NJ.

Ridgely, R.S. and Gaulin, S.J. (1980) The birds of Finca Merenberg, Huila Department, Colombia. *Condor*, **82**, 379–91.

Robbins, C.S., Droege, S. and Sauer, J.R. (1989) Monitoring bird populations with Breeding Bird Survey and atlas data. *Annales Zoologici Fennici*, **26**, 297–304.

Roberson, D. (1980) *Rare Birds of the West Coast of North America*, Woodcock, Pacific Grove, CA.

Roberts, C.M., Dawson Shepherd, A.R. and Ormond, R.F.G. (1992) Large-scale variation in assemblage structure of Red Sea butterflyfishes and angelfishes. *Journal of Biogeography*, **19**, 239–50.

Roberts, P.R. and Oosting, H.J. (1958) Responses of venus fly trap (*Dionaea muscipula*) to factors involved in its endemism. *Ecological Monographs*, **28**, 193–218.

Robey, E.H., Smith, H.D. and Belk, M.C. (1987) Niche pattern in a Great Basin rodent fauna. *Great Basin Naturalist*, **47**, 488–96.

Rogers, D.J. and Randolph, S.E. (1991) Mortality rates and population density of tsetse flies correlated with satellite imagery. *Nature*, **351**, 739–41.

Rogers, R.S. (1983) Annual variability in community organization of forest herbs: effect of an extremely warm and dry early spring. *Ecology*, **64**, 1086–91.

Room, P.M., Harley, K.L.S., Forno, I.W. and Sands, D.P.A. (1981) Successful biological control of the floating weed salvinia. *Nature*, **294**, 78–80.

Root, T. (1988) *Atlas of Wintering North American Birds: An Analysis of Christmas Bird Count Data*, University of Chicago Press, Chicago, IL.

Root, T. (1991) Positive correlation between range size and body size: a possible mechanism. *Acta XX Congressus Internationalis Ornithologici*, 817–25.

Rosenzweig, M.L. (1975) On continental steady states of species diversity, in *Ecology and Evolution of Communities* (eds M.L. Cody and J.M. Diamond), Harvard University Press, Cambridge, MA, pp. 124–40.

Roubik, D.W. and Ackerman, J.D. (1987) Long-term ecology of euglossine orchid-bees (Apidae: Euglossini) in Panama. *Oecologia (Berl.)*, **73**, 321–33.

Roy, J., Navas, M.L. and Sonie, L. (1991) Invasion by annual brome grasses: a case study challenging the homocline approach to invasions, in *Biogeography of Mediterranean Invasions* (eds R.H. Groves and F. Di Castri), Cambridge University Press, Cambridge, pp. 207–24.

Russell, M.P. and Lindberg, D.R. (1988a) Real and random patterns associated with molluscan spatial and temporal distributions. *Paleobiology*, **14**, 322–30.

Russell, M.P. and Lindberg, D.R. (1988b) Estimates of species duration. *Science (Wash.)*, **240**, 969.

Ryti, R.T. (1992) Effect of the focal taxon on the selection of nature reserves. *Ecological Applications*, **2**, 404–10.

Sale, P.F. (1988) Perception, pattern, chance and the structure of reef fish communities. *Environmental Biology of Fishes*, **21**, 3–15.

Samways, M.J. (1990) Species temporal variability: epigaeic ant assemblages and management for abundance and scarcity. *Oecologia (Berl.)*, **84**, 482–90.

Sayer, J.A. (1982) The pattern of the decline of the korrigum *Damaliscus lunatus* in West Africa. *Biological Conservation*, **23**, 95–110.

Schall, J.J. and Pianka, E.R. (1978) Geographical trends in numbers of species. *Science (Wash.)*, **201**, 679–86.

Schmidt, J.O. and Buchmann, S.L. (1986) Are mutillids scarce? (Hymenoptera: Mutillidae). *Pan-Pacific Entomologist*, **62**, 103–4.

Schmidt, K.P. (1950) The concept of geographic range, with illustrations from amphibians and reptiles. *The Texas Journal of Science*, **3**, 326–34.

Schoener, T.W. (1987) The geographical distribution of rarity. *Oecologia (Berl.)*, **74**, 161–73.

Schoener, T.W. (1989) The ecological niche, in *Ecological Concepts: The Contribution of Ecology to an Understanding of The Natural World* (ed. J.M. Cherrett), Blackwell Scientific, Oxford, pp. 79–113.

Schoener, T.W. (1990) The geographical distribution of rarity: misinterpretation of atlas methods affects some empirical conclusions. *Oecologia (Berl.)*, **82**, 567–8.

Scholtz, C.H. and Chown, S.L. (1993) Insect conservation and extensive agriculture: the savanna of southern Africa, in *Perspectives on Insect Conservation* (eds. K.J. Gaston, T.R. New and M.J. Samways), Intercept, Andover, in press.

Schonewald-Cox, C., Azari, R. and Blume, S. (1991) Scale, variable density, and conservation planning for mammalian carnivores. *Conservation Biology*, **5**, 491–5.

Schonewald-Cox, C. and Buechner, M. (1991) Housing viable populations in protected habitats: the value of a coarse-grained geographic analysis of density patterns and available habitat, in *Species Conservation: A Population–Biological Approach* (eds A. Seitz and V. Loeschcke), Birkhauser Verlag, Basel, pp. 213–26.

Schonewald-Cox, C.M., Chambers, S.M., MacBryde, B. and Thomas, L. (eds) (1983) *Genetics and Conservation: A Reference for Managing Wild Animal and Plant Populations*, Benjamin/Cummings, Meulo Park, CA.

Scott, J.M. and Kepler, C.B. (1985) Distribution and abundance of Hawaiian native birds: a status report. *Bird Conservation 2* (ed. S.A. Temple), University of Wisconsin Press, Madison, pp. 43–70.

Scott, J.M., Csuti, B., Jacobi, J.D. and Estes, J.E. (1987) Species richness: a geographic approach to protecting future biological biodiversity. *BioScience*, **37**, 782–8.

Scott, P., Burton, J.A. and Fitter, R. (1987) Red Data Books: the historical background, in *The Road to Extinction* (eds R. Fitter and M. Fitter), IUCN/UNEP, Gland, pp. 1–5.

Seagle, S.W. and McCracken, G.F. (1986) Species abundance, niche position, and niche breadth for five terrestrial animal assemblages. *Ecology*, **67**, 816–18.

Seber, G.A.F. (1982) *The Estimation of Animal Abundance and Related Parameters* (2nd edn), Macmillan, New York.

Shenbrot, G.I., Rogovin, K.A. and Surov, A.V. (1991) Comparative analysis of spatial organization of desert lizard communities in Middle Asia and Mexico. *Oikos*, **41**, 157–68.

Shkedy, Y. and Safriel, U.N. (1992) Niche breadth of two lark species in the desert and the size of their geographical ranges. *Ornis Scandinavica*, **23**, 89–95.

Shmida, A. and Wilson, M.V. (1985) Biological determinants of species diversity. *Journal of Biogeography*, **12**, 1–20.

Shugart, H.H. and Patten, B.C. (1972) Niche quantification and the concept of niche pattern, in *Systems Analysis and Simulation Ecology* (ed. B.C. Patten), Academic Press, New York, pp. 283–327.

Siegel, S. (1956) *Nonparametric Statistics for the Behavioural Sciences,* McGraw-Hill, New York.

Slater, F.M., Curry, P. and Chadwell, C. (1987) A practical approach to the evaluation of the conservation status of vegetation in river corridors in Wales. *Biological Conservation*, **40**, 53–68.

Smith, C.R. (ed.) (1990) *Handbook for Atlasing North American Breeding Birds*, Vermont Institute of Natural Science, Woodstock, Vermont.

Smith, E.P. (1982) Niche breadth, resource availability, and inference. *Ecology*, **63**, 1675–81.

Söderström, L. (1989) Regional distribution patterns of bryophyte species on spruce logs in northern Sweden. *The Bryologist*, **92**, 349–55.

Solem, A. (1984) A world model of land snail diversity and abundance, in *World-wide Snails: Biogeographical Studies on Non-marine Mollusca* (eds A. Solem and A.C. Van Bruggen), E.J. Brill/Dr W. Backhuys, Leiden, pp. 6–22.

Soulé, M.E. (1983) What do we really know about extinction? in *Genetics and Conservation: A Reference for Managing Wild Animal and Plant Populations* (eds C.M. Schonewald-Cox, S.M. Chambers, B. MacBryde and L. Thomas), Benjamin/Cummings, Meulo Park, CA, pp. 111–24.

Soulé, M.E. (1986) Patterns of diversity and rarity: their implications for conservation, in *Conservation Biology: the Science of Scarcity and Diversity* (ed. M.E. Soulé), Sinauer Associates, Sunderland, MA, pp. 117–21.

Soulé, M.E. (1987) Where do we go from here? in *Viable Populations for Conservation* (ed. M.E. Soulé), Cambridge University Press, Cambridge, pp. 175–83.

Soulé, M.E. (1991) Conservation: tactics for a constant crisis. *Science (Wash.)*, **253**, 744–50.

Soulé, M.E., Bolger, D.T., Alberts, A.C. *et al.* (1988) Reconstructed dynamics of rapid extinctions of chaparral-requiring birds in urban habitat islands. *Conservation Biology*, **2**, 75–92.

Southwood, T.R.E. (1978) *Ecological Methods*, Chapman & Hall, London.

Southwood, T.R.E. (1981) Ecological aspects of insect migration, in *Animal Migration* (ed. D.J. Aidley), Cambridge University Press, Cambridge, pp. 197–208.

Spellerberg, I.F. (1981) *Ecological Evaluation for Conservation*, Edward Arnold, London.

Spellerberg, I.F. (1991) Biogeographical basis of conservation, in *The Scientific Management of Temperate Communities for Conservation* (eds I.F. Spellerberg, F.B. Goldsmith and M.G. Morris), Blackwell Scientific, Oxford, pp. 293–322.

Spellerberg, I.F. (1992) *Evaluation and Assessment for Conservation*, Chapman & Hall, London.

Spitzer, K. and Lepš, J. (1988) Determinants of temporal variation in moth abundance. *Oikos*, **53**, 31–6.

Stacey, P.B. and Taper, M. (1992) Environmental variation and the persistence of small populations. *Ecological Applications*, **2**, 18–29.

Stanley, S.M. (1986) Population size, extinction, and speciation: the fission effect in Neogene Bivalvia. *Paleobiology*, **12**, 89–110.

Stebbins, G.L. (1978a) Why are there so many rare plants in California? 1. Environmental factors. *Fremontia*, **5**, 6–10.

Stebbins, G.L. (1978b) Why are there so many rare plants in California? 2. Youth and age of species. *Fremontia*, **6**, 17–20.

Stebbins, G.L. and Major, J. (1965) Endemism and speciation in the California flora. *Ecological Monographs*, **35**, 1–35.

Stemberger, R.S. and Gilbert, J.J. (1985) Body size, food concentration, and population growth in planktonic rotifers. *Ecology*, **66**, 1151–9.

Stevens, G.C. (1986) Dissection of the species–area relationship among wood-boring insects and their host plants. *American Naturalist*, **128**, 35–46.

Stevens, G.C. (1989) The latitudinal gradient in geographical range: how so many species coexist in the tropics. *American Naturalist*, **133**, 240–56.

Stevens, G. (1992a) Spilling over the competitive limits to species coexistence, in *Systematics, Ecology, and the Biodiversity Crisis* (ed. N. Eldredge), Columbia University Press, New York, pp. 40–58.

Stevens, G.C. (1992b) The elevational gradient in altitudinal range: an extension of Rapoport's latitudinal rule to altitude. *American Naturalist*, **140**, 893–911.

Stiles, F.G. (1983) Checklist of birds, in *Costa Rican Natural History* (ed. D.H. Janzen), University of Chicago Press, Chicago, pp. 530–44.

Strayer, D.L. (1991) Projected distribution of the zebra mussel, *Dreissena polymorpha*, in North America. *Canadian Journal of Fisheries and Aquatic Sciences*, **48**, 1389–95.

Strong, D.R., Lawton, J.H. and Southwood, T.R.E. (1984) *Insects on Plants: Community Patterns and Mechanisms*, Blackwell Scientific, Oxford.

Stubbs, A.E. (1982) Conservation and the future for the field entomologist. *Proceedings and Transactions of the British Entomological and Natural History Society*, **15**, 55–67.

Sugihara, G. (1980) Minimal community structure: an explanation of species abundance patterns. *American Naturalist*, **116**, 770–87.

Sukopp, H. and Trautmann, W. (1981) Causes of the decline of threatened plants in the Federal Republic of Germany, in *The Biological Aspects of Rare Plant Conservation* (ed. H. Synge), Wiley, New York, pp. 113–16.

Sutherland, W.J. and Baillie, S.R. (1993) Patterns in the distribution, abundance and variation of bird populations. *Ibis*, **135**, 209–10.

Svensson, B.W. (1992) Changes in occupancy, niche breadth and abundance of three *Gyrinus* species as their respective range limits are approached. *Oikos*, **63**, 147–56.

Taylor, K., Fuller, R.J. and Lack, P.C. (eds) (1985) *Bird Census and Atlas Studies*, British Trust for Ornithology, Tring.

Taylor, L.R., Woiwod, I.P. and Perry, J.N. (1978) The density dependence of spatial behaviour and the rarity of randomness. *Journal of Animal Ecology*, **47**, 383–406.

Taylor, R.A.J. and Taylor, L.R. (1979) A behavioural model for the evolution of spatial dynamics, in *Population Dynamics* (eds R.M. Anderson, B.D. Turner and L.R. Taylor), Blackwell Scientific, Oxford, pp. 1–27.

Terborgh, J. (1988) The big things that run the world – A sequel to E.O. Wilson. *Conservation Biology*, **2**, 402–3.

Terborgh, J. (1989) *Where Have All the Birds Gone?*, Princeton University Press, Princeton, NJ.

Terborgh, J., Robinson, S.K., Parker, T.A. III *et al.* (1990) Structure and organization of an Amazonian forest bird community. *Ecological Monographs*, **60**, 213–38.

Terborgh, J. and Winter, B. (1980) Some causes of extinction, in *Conservation Biology: An Evolutionary–Ecological Perspective* (eds M.E. Soulé and B.A. Wilcox), Sinauer Associates, Sunderland, MA, pp. 119–33.

Thomas, B.T. (1979) The birds of a ranch in the Venezuelan llanos, in *Vertebrate Ecology in the Northern Neotropics* (ed. J.F. Eisenberg), Smithsonian Institution, Washington, DC, pp. 213–32.

Thomas, C.D. (1991) Habitat use and geographic ranges of butterflies from the wet lowlands of Costa Rica. *Biological Conservation*, **55**, 269–81.

Thomas, C.D. and Mallorie, H.C. (1985a) On the altitudes of Moroccan butterflies. *Entomologist's Monthly Magazine*, **121**, 253–6.

Thomas, C.D. and Mallorie, H.C. (1985b) Rarity, species richness and conservation: butterflies of the Atlas mountains in Morocco. *Biological Conservation*, **33**, 95–117.

Thomas, J.A. (1991) Rare species conservation: case studies of European butterflies, in *The Scientific Management of Temperate Communities for Conservation* (eds I.F. Spellerberg, F.B. Goldsmith and M.G. Morris), Blackwell Scientific, Oxford, pp. 149–97.

Tokeshi, M. (1990) Niche apportionment or random assortment: species abundance patterns revisited. *Journal of Animal Ecology*, **59**, 1129–46.

Tonn, W.M., Magnuson, J.J., Rask, M. and Toivonen, J. (1990) Intercontinental comparison of small-lake fish assemblages: the balance between local and regional processes. *American Naturalist*, **136**, 345–75.

Townsend, C.R., Hildrew, A.G. and Schofield, K. (1987) Persistence of stream invertebrate communities in relation to environmental variability. *Journal of Animal Ecology*, **56**, 597–613.

Tracy, C.R. and George, T.L. (1992) On the determinants of extinction. *American Naturalist*, **139**, 102–22.

Turin, H. and den Boer, P.J. (1988) Changes in the distribution of carabid beetles in The Netherlands since 1880. II. Isolation of habitats and long-term time trends in the occurrence of carabid species with different powers of dispersal (Coleoptera, Carabidae). *Biological Conservation*, **44**, 179–200.

Udvardy, M.D.F. (1969) *Dynamic Zoogeography: with Special Reference to Land Animals*, Van Nostrand Reinhold, New York.

Urban, D.L. and Smith, T.M. (1989) Microhabitat pattern and the structure of forest bird communities. *American Naturalist*, **133**, 811–29.

Usher, M.B. (1986a) Wildlife conservation evaluation: attributes, criteria and values, in *Wildlife Conservation Evaluation* (ed. M.B. Usher), Chapman & Hall, London, pp. 3–44.

Usher, M.B. (1986b) Insect conservation: the relevance of population and community ecology and of biogeography, in *Proceedings of the 3rd European Congress of Entomology. Part 3* (ed. H.H.W. Velthuis), Nederlandse Entomologische Vereniging, Amsterdam, pp. 387–98.

van der Ploeg, S.W.F. (1986) Wildlife conservation evaluation in the Netherlands: a controversial issue in a small country, in *Wildlife Conservation Evaluation* (ed. M.B. Usher), Chapman & Hall, London, pp. 161–80.

van der Ploeg, S.W.F. and Vlijm, L. (1978) Ecological evaluation, nature conservation and land use planning with particular reference to methods used in The Netherlands. *Biological Conservation*, **14**, 197–221.

van Swaay, C.A.M. (1990) An assessment of the changes in butterfly abundance in The Netherlands during the 20th century. *Biological Conservation*, **52**, 287–302.

Van Valen, L. (1973a) A new evolutionary law. *Evolutionary Theory*, **1**, 1–30.

Van Valen, L. (1973b) Body size and numbers of plants and animals. *Evolution*, **27**, 27–35.

Verkaar, H.J. (1990) Corridors as a tool for plant species conservation? in *Species Dispersal in Agricultural Habitats* (eds R.G.H. Bunce and D.C. Howard), Belhaven Press, London, pp. 82–97.

Verner, J. (1985) Assessment of counting techniques, in *Current Ornithology Vol. 2*, (ed. R.F. Johnston), pp. 247–302.

Walker, B.H. (1992) Biodiversity and ecological redundancy. *Conservation Biology*, **6**, 18–23.

Walker, P.A. (1990) Modelling wildlife distributions using a geographic information system: kangaroos in relation to climate. *Journal of Biogeography*, **17**, 279–89.

Walter, H.S. (1990) Small viable population: the red-tailed hawk of Socorro Island. *Conservation Biology*, **4**, 441–3.

Ward, J.P. Jr., Franklin, A.B. and Gutierrez, R.J. (1991) Using search time and regression to estimate abundance of territorial spotted owls. *Ecological Applications*, **1**, 207–14.

Ward, S.A., Sunderland, K.D., Chambers, R.J. and Dixon, A.F.G. (1986) The use of incidence counts for estimation of cereal aphid populations. 3. Population development and the incidence–density relation. *Netherlands Journal of Plant Pathology*, **92**, 175–83.

Wasserman, S.S. and Mitter, C. (1978) The relationship of body size to breadth of diet in some Lepidoptera. *Ecological Entomology*, **3**, 155–60.

Webb, T. III (1987) The appearance and disappearance of major vegetational assemblages: long-term vegetational dynamics in eastern North America. *Vegetatio*, **69**, 177–87.

Werner, J.K. (1982) Anurans, in *CRC Handbook of Census Methods for Terrestrial Vertebrates* (ed. D.E. Davis), CRC Press, Boca Raton, FL, pp. 9–10.

Wheeler, B.D. (1988) Species richness, species rarity and conservation evaluation of rich-fen vegetation in lowland England and Wales. *Journal of Applied Ecology*, **25**, 331–52.

White, G. (1788) *The Natural History of Selborne*, reprinted 1981, Penguin, Harmondsworth.

White, P.S. and Miller, R.I. (1988) Topographic models of vascular plant richness in the southern Appalachian high peaks. *Journal of Ecology*, **76**, 192–9.

White, P.S., Miller, R.I. and Ramseur, G.S. (1984) The species–area relationship of the southern Appalachian high peaks: Vascular plant richness and rare plant distributions. *Castanea*, **49**, 47–61.

Whiteman, P. and Millington, R. (1991) The British list and rare birds in the eighties. *Birding World*, **3**, 429–34.

Wiens, D., Nickrent, D.L., Davern, C.I. and Calvin, C.L. (1989) Developmental failure and loss of reproductive capacity in the rare palaeoendemic shrub *Dedeckera eurekensis. Nature*, **338**, 65–7.

Wiens, J.A. (1989) *The Ecology of Bird Communities, Vol. 1. Foundations and Patterns*, Cambridge University Press, Cambridge.

Wilcove, D.S. and Terborgh, J.W. (1984) Patterns of population decline in birds. *American Birds*, **38**, 10–13.

Williams, C.B. (1950) The application of the logarithmic series to the frequency of occurrence of plant species in quadrats. *Journal of Ecology*, **38**, 107–38.

Williams, C.B. (1960) The range and pattern of insect abundance. *American Naturalist*, **94**, 137–51.

Williams, C.B. (1964) *Patterns in the Balance of Nature*, Academic Press, London.

Williams, G.R. and Given, D.R. (1981) *The Red Data Book of New Zealand: Rare and Endangered Species of Endemic Terrestrial Vertebrates and Vascular Plants*, Nature Conservation Council, Wellington.

Williams, J.E. (1991) Biogeographic patterns of three sub-alpine eucalypts in south-east Australia with special reference to *Eucalyptus pauciflora* Sieb. ex Spreng. *Journal of Biogeography*, **18**, 223–30.

Williams, P.H. (1988) Habitat use by bumble bees (*Bombus* spp.). *Ecological Entomology*, **13**, 223–37.

Williams, P.H. (1991) The bumble bees of the Kashmir Himalaya (Hymenoptera: Apidae, Bombini). *Bulletin of the British Museum Natural History (Entomology)*, **60**, 1–204.

Williams, P.H. and Gaston, K.J. (1994) Measuring more of biodiversity: can higher-taxon richness predict wholesale species richness? *Biological Conservation*, in press.

Williamson, M. (1989a) Natural extinction on islands. *Philosophical Transactions of the Royal Society of London (Series B)*, **325**, 457–68.

Williamson, M. (1989b) Mathematical models of invasion, in *Biological Invasions: A Global Perspective* (eds J.A. Drake, H.A. Mooney, F. di Castri *et al.*), Wiley, Chichester, pp. 329–50.

Williamson, M.H. (1987) Are communities ever stable? in *Colonisation, Succession and Stability* (eds A.J. Gray, M.J. Crawley and P.J. Edwards), Blackwell Scientific, Oxford, pp. 352–71.

Williamson, M.H. and Lawton, J.H. (1991) Fractal geometry of ecological habitats, in *Habitat Structure: the Physical Arrangement of Objects in Space* (eds S.S. Bell, E.D. McCoy and H.R. Mushinsky), Chapman & Hall, London, pp. 69–86.

Willis, J.C. (1922) *Age and Area: A Study in Geographical Distribution and Origin of Species*, Cambridge University Press, Cambridge.

Wilson, E.O. (1987) Causes of ecological success: the case of the ants. *Journal of Animal Ecology*, **56**, 1–9.

Wilson, L.T. and Room, P.M. (1983) Clumping patterns of fruit and arthropods in cotton, with implications for binomial sampling. *Environmental Entomology*, **12**, 50–4.

Wolda, H. (1981) Similarity indices, sample size and diversity. *Oecologia (Berl.)*, **50**, 296–302.

Wolfheim, J.H. (1983) *Primates of the World: Distribution, Abundance and Conservation*, Harwood Academic Publishers, Chur.

Woodward, F.I. (1987) *Climate and Plant Distribution*, Cambridge University Press, Cambridge.

Wright, D.F. (1977) A site evaluation scheme for use in the assessment of potential nature reserves. *Biological Conservation*, **11**, 293–305.

Wright, D.H. (1983) Species–energy theory: an extension of species–area theory. *Oikos*, **41**, 496–506.

Wright, D.H. (1991) Correlations between incidence and abundance are expected by chance. *Journal of Biogeography*, **18**, 463–6.

Xia, X. and Boonstra, R. (1992) Measuring temporal variability of population density: a critique. *American Naturalist*, **140**, 883–92.

Yamamura, K. (1990) Sampling scale dependence of Taylor's power law. *Oikos*, **59**, 121–5.

Youtie, B.A. (1987) Herbivorous and parasitic insect guilds associated with Great Basin wildrye (*Elymus cinereus*) in southern Idaho. *Great Basin Naturalist*, **47**, 644–51.

Author index

Subject index